Service Industry Databook

B. Elango

Service Industry Databook

Understanding and Analyzing Sector Specific Data Across 15 Nations

 Springer

B. Elango
College of Business
Illinois State University
Normal, IL
USA

ISBN 978-3-319-36593-0   ISBN 978-3-319-19111-9   (eBook)
DOI 10.1007/978-3-319-19111-9

Springer Cham Heidelberg New York Dordrecht London

Printed on acid-free paper

Springer International Publishing AG Switzerland is part of Springer Science+Business Media
(www.springer.com)

Dedicated to my parents,
Late Sri. S. Balasubramanian
and Smt. Subbulashmi Balasubramanian,
in recognition of their patience and support.

Preface

Although I have published over 60 articles and been recognized by my peers for my work, I have always felt the need to write a book and had this dream in the back of my mind. Over the past decade, several of my research projects have focused on the service industry. While I was able to achieve my end results with each of these projects by publishing them as articles in top-tier academic journals, one of their most challenging aspects was the difficulty in locating data on the service sector.

After wrapping up my last research project in this area, it occurred to me that a book which served as an information primer on the service sector would be a valuable asset to researchers like myself, industry practitioners, and public policymakers. Additionally, I was convinced that there was no other comparable alternative resource, as I have been searching for one like it for many years. I thought that if there is a lack of data for the US market itself (a market for which more information is typically available than most), it would likely be much more difficult to locate similar data for other markets. This led me to envision a reference book to fill the (admittedly narrow) niche in the market. However, there was one thing I did not recognize when I committed myself to this project. Despite this project's narrow niche, its scope is necessarily wide, and I soon discovered that it would be impossible to claim to be any kind of expert on all nations and industry segments. I do understand that much of the information presented in the book can be replicated by someone else, provided one has the background knowledge and access to the same information. However, I envision it helping many individuals by saving time, as it is a ready store of information and should serve as a quick reference on the service sector. Thank you for reading this book, and I hope it serves you well by presenting the information you seek.

Normal, IL, USA   Dr. B. Elango
elango@ilstu.edu;
mrelango@hotmail.com

Contents

List of Figures

List of Tables

Introduction

1.1 Overview

The importance of services to the world economy has been well-recognized. The service sector is responsible for two-thirds of world output and about one-fifth of total trade worldwide. Globally, service output, which was valued at US$ 1,693,364 million in 1970, has risen to US$ 45,323,667 million in 2012, an increase of 2577 % over 42 years (61 % average growth rate). Figure 1.1 graphically presents the combined output of the value of services across developing economies, transition economies and developed economies. Figure 1.2 graphically illustrates the growth of the service sector across developing economies, transition economies and developed economies. In recent years (2003–2012), the growth of the service sector in developing and transition economies has been much higher than in developed countries. For instance, in the decade spanning 2003–2014, the service sector grew at rate of 13 % for developing economies, 19 % for transition economies and 5.4 % for developed economies. Among the three economy types, year-to-year growth shows the greatest variation for transition economies and developed economies. Figure 1.3 graphically illustrates value of imports and exports for the service sector. Overall, imports have grown starting at US$ 139,580 million in 1980 and ending at US$ 1,696,750 million in 2013 (1115 % increase), while exports have grown starting at US$ 395,660 million in 1980 and ending at US$ 4,720,180 million in 2013 (1092 % increase). The absolute numbers and the consistent growth pattern over the period indicate the importance of this sector, which represents about 44 % of world employment. Services represent about 74 % of employment in developed countries and 37 % in developing countries. It is also estimated that the service sector received about $570 billion in FDI investments, which is about 40 % of global foreign direct investment (UNCTAD 2013).

© Springer International Publishing Switzerland 2015   1
B. Elango, *Service Industry Databook: Understanding and Analyzing Sector Specific Data Across 15 Nations*, DOI 10.1007/978-3-319-19111-9_1

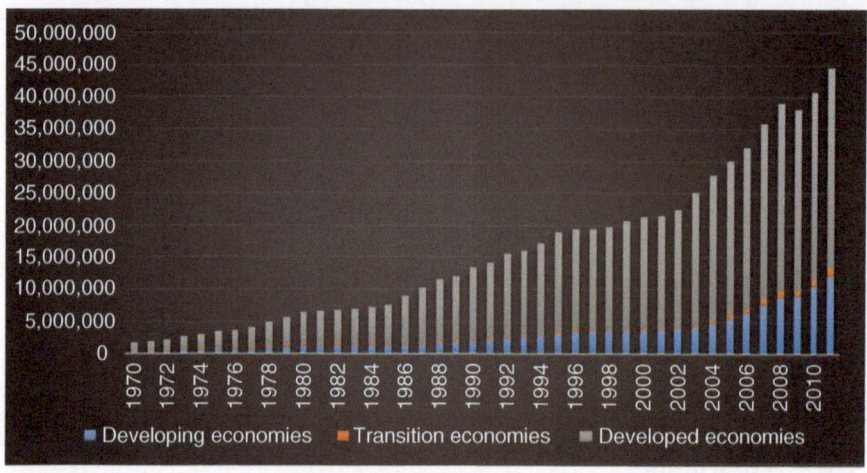

Fig. 1.1 Service sector output for the years 1970–2012 (Millions of US Dollars) (Data source: United Nations Conference on Trade and Development (UNCTAD))

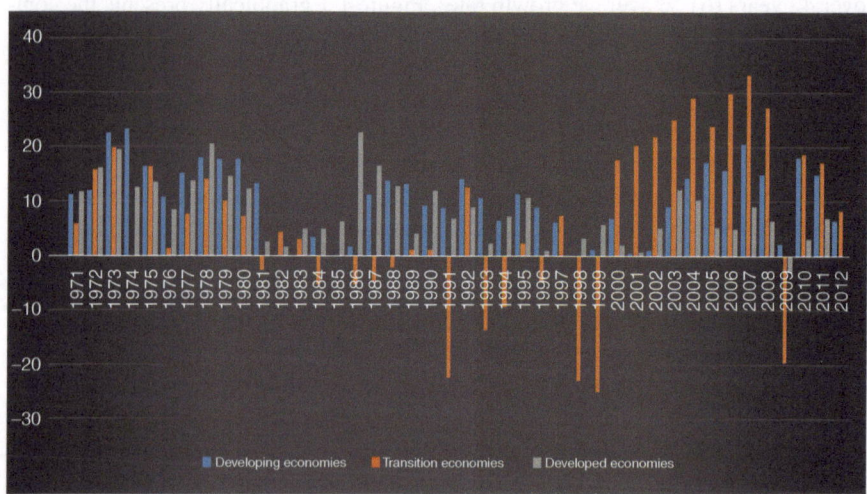

Fig. 1.2 Growth rate of service sector for the years 1970–2012 (Percentage) (Data source: United Nations Conference on Trade and Development (UNCTAD))

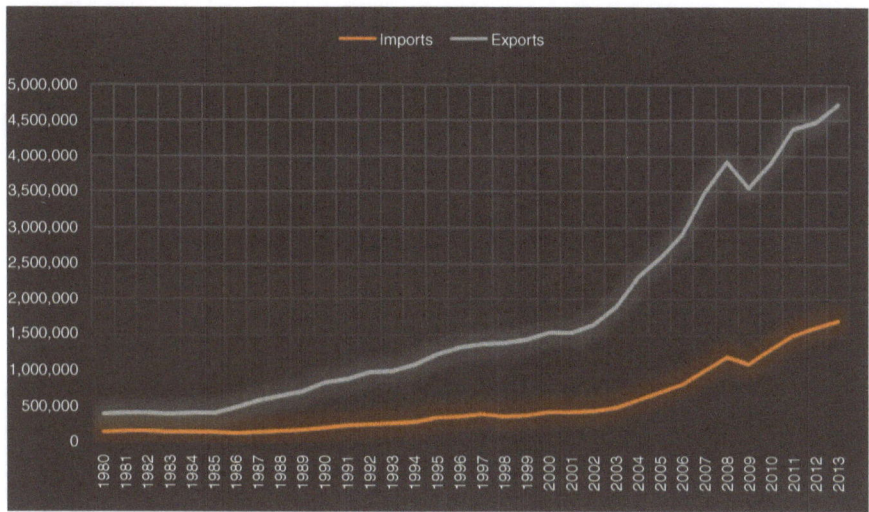

Fig. 1.3 Values of exports and imports for the years 1980–2013 (Millions of US Dollars) (Data source: United Nations Conference on Trade and Development (UNCTAD))

1.2   Need for This Book

While the trends presented here document the importance of the service sector, finding information on it is not an easy task. The service sector can be divided into many industry segments, examples of which include telecommunication, construction, transportation, insurance, financial services, IT services, accounting & legal services, advertising services, R&D services, engineering services, education, personal & cultural services, and health services. However, finding a ready source of data for the service industry seems to be a task in futility. For instance, using common search engines or a review of online book stores' extensive collections typically leads to nothing fruitful. This glaring hole is obvious, as this is not true for the manufacturing sector. A search typically offers many leads to specific sources or publications. Additionally, many academic research publications exist which focus on the various manufacturing industry sectors. This is in contrast to academic publications on the service sector, which are either based on a single sector or on aggregate classifications (e.g., capital-intensive versus knowledge-intensive industries or hard versus soft service). As a researcher, the author recognizes why this may be the case, given the lack of specific sources which provide industry information on the service sector to do research. This book seeks to serve this gap in the market.

1.3   Structure of the Book

This book is organized into six chapters inclusive of this introductory chapter. Chapter 2 presents various sources available to a researcher interested in gathering empirical data on the service sector. It introduces the reader to finding key sources of data and understanding the advantages and disadvantages of data from these sources. It will serve any researcher seeking to collect and conduct research on a

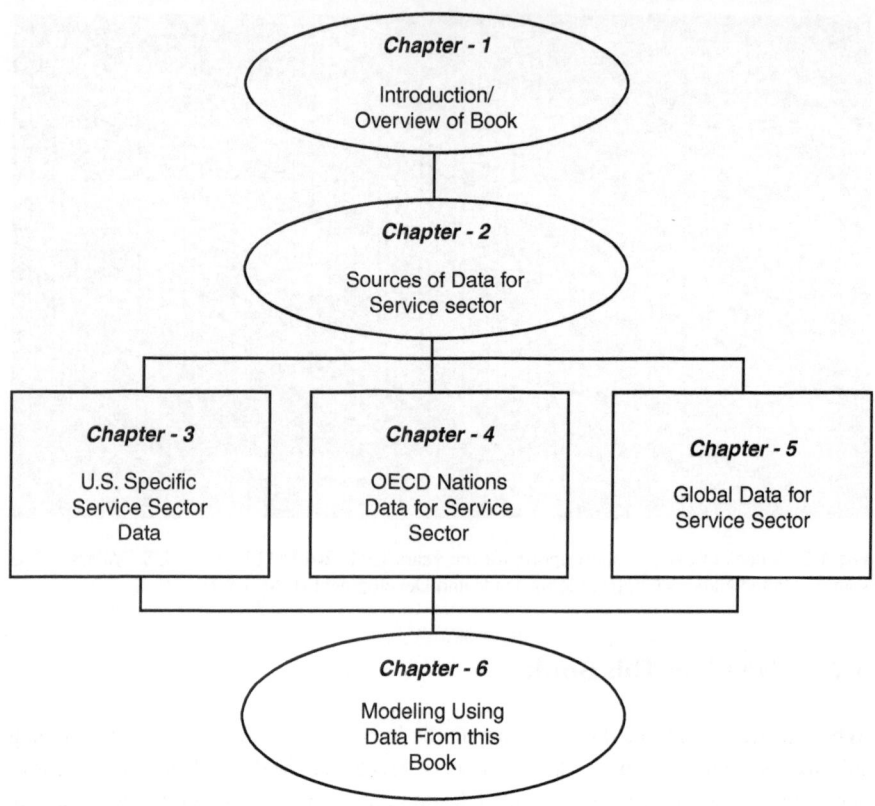

Fig. 1.4 Structure of the book

specific service on their own. Chapter 3 focuses exclusively on service sector indus-
tries in the United States. It provides data on U.S. Service Exports (1999–2013);
U.S. Service Imports (1999–2013); U.S. Export Ratio; U.S. Import Ratio; Foreign
Investment Ratio (1997–2013); Asset Intensity (1997–2013); Labor Ratio (1998–
2013); Profit Margin Before Tax (1998–2012); Industry Output (2000–2013), with
notable missing values for the later years. Chapter 4 provides data on fifteen (15)
OECD countries (i.e., Austria; Belgium; Czech Republic; Denmark; Finland;
France; Germany; Hungary; Italy; Korea; Netherlands; Norway; Slovenia; Sweden;
and United States) with Export Ratio, Import Ratio, Profit Ratio, and DGM Ratio
for the years 2000–2010. United States is covered in both Chapters 3 and 4, as data
was available from two different sources. Chapter 3 uses data from the
U.S. Department of Commerce to compute these figures, whereas Chapter 4 uses
data from the Organisation for Economic Cooperation and Development (OECD).
As a result, the correspondence between the numbers reported may vary. Chapter 5
uses information on public companies from Standard & Poor's GlobalVantage
Database and computes aggregate ratios such as selling and administrative ratio,
leverage, gross profits, and industry concentration of public companies for the years
2004–2013. Chapter 6, the conclusion, presents the author's empirical models using
data from sources discussed in this book to test the factors driving profitability of
the service industry. Figure 1.4 provides an illustration of the structure of the book.

1.4   Measures Provided in This Book

The various data variables and measures provided are elaborated briefly below.

1.4.1   Asset Intensity

Asset Intensity is measured by dividing fixed assets by outputs of the industry. This variable is used to capture the extent of assets needed to produce services. Industries which have a higher asset intensity typically require higher investments in assets relative to the sales it can generate. This variable has also been referred to as physical capital intensity (Elango 2005). Additionally, firms entering industries with high asset intensity should recognize that their ability to exit this industry without a good market for buyers could potentially lead to very expensive write-offs. An asset intensity score of 245 denotes that to generate one unit output of service, an investment to the tune of 2.45 times is needed. This book provides asset intensity information for the United States across industries in Chapter 3.

1.4.2   Concentration

Concentration refers to the size distribution of firms in an industry (Curry and George 1983. Concentration is one of the most critical elements in an industry's market structure. In a concentrated industry, fewer firms control a large market share. The common assumption is that these firms would be able to use their market power to their advantage and maximize profits for two reasons. First, these firms typically need not compete intensively to gain market share. Second, these firms would be able to use their market position to prevent new rivals from getting a hold in their market (Bain 1959). Empirical evidence supports this assertion (Weiss 1979; Dalton and Penn 1976) even though there have been a few instances where contrary evidence has been reported (Attaran and Saghafi 1988). This variable is considered critical and is used extensively in managerial practice and antitrust law applications to judge the extent of competition in an industry. Concentration has been measured several ways, namely, using Herfindahl ratio, C4, C8, C20, C50, Hirschman Index, Entropy coefficient and Gini coefficient. This book measures concentration ratios as C4 and C8 based on Scherer (1980) as the percentage of total industry sales contributed by the four largest public firms or eight largest public firms, respectively. For example, in the instance where the top four firms account for all the sales in an industry, the score would be 1 (highest score possible), or in the case where the top four firms control 24 % of the market, the score would be .24. This book reports concentration ratios (C4 and C8) globally across industries in Chapter 5.

1.4.3   Domestic Global Market Ratio (DGM Ratio)

Domestic Global Market Ratio compares the size of the domestic market relative to the global market in a particular industry. The size of the domestic market sets a cap on local firm growth and thereby impacts a firm's prospects and potential. In some instances a small domestic market (relative to the global size of the market) may force pioneering local firms to seek other markets, while on the other hand a relatively large domestic market may encourage many foreign rivals to seek markets locally (Shapiro and Taylor 1990). This variable, much ignored until recently, has been found to be a critical driver of a firm's motivation to internationalize its operations (Elango 1998). Domestic Global Market Ratio is operationalized as the total domestic market sales divided by total global sales for a particular industry. Therefore, for instance, where a country represents 5 % of the total industry global sales, the DGM Ratio would be 5. In this book, the Domestic Global Market Ratio for OECD nations across industries can be found in Chapter 5.

1.4.4　Exports and Export Ratio

Exports refers to the value of services that have been transferred from the nation of interest (across national borders) to a foreign country. The extent of the exports serve as a proxy for the strength of a particular industry relative to other countries. The underlying logic for this assumption is that firms in industries which do not have relative comparative advantages will not succeed in foreign markets. Exports by local firms offer several advantages. First, exporting of industry output results in increased market space for all firms, as it is limited by local market size. Typically, export activity allows for greater firm variation and growth of firms in an industry. Second, it enhances the opportunity to learn and adapt new ideas from foreign markets to implement them locally. While Exports represents the aggregate value of services transferred outside a country's borders, it does not offer a relative assessment as to what extent it represents the level of value activity in a particular industry. Export Ratio presents this information by measuring the extent of exports in an industry. An industry which exports 20 % of its output will have a score of .2. This book reports the exports for the United States in Chapter 3 and Export Ratio for the United States and OECD nations across industries in Chapter 3.

1.4.5　Gross Profit Ratio

Gross Profit Ratio refers to the gross profit margins made by a firm on its sales and is measured by dividing gross profit by sales revenue, while gross profit is measured as the sales revenue minus the cost of goods sold. Gross Profit Ratio is an important number, as it is quite indicative of the extent to which a firm can make profits (after expensing other costs), since the cost of production of a service has already been accounted for (Elango and Sethi 2007). Therefore, typically, Gross Profit Ratio signals the degree of attractiveness of an industry. A firm whose gross profits represents 30 % of the sales revenue will have a Gross Profit Ratio of .3. This book provides Gross Profit Ratio information on across industries for OECD nations in Chapter 4 and globally across industries in Chapter 5. Unlike in Chapter 4, it should be noted that in Chapter 5, this ratio is expressed as a percentage.

1.4.6   Foreign Investment Ratio

Foreign Investment Ratio represents the extent of investment made by foreign (non-domiciled) firms in a particular service sector. This variable is measured by dividing the foreign direct investment position on a historical cost basis by the value of the private fixed assets on a historical cost basis in an industry. This variable serves as a good proxy of foreign competition in service industries, as many services typically cannot be inventoried. An industry with a Foreign Investment Ratio of 21.24 indicates that foreign firms control about 21.24 % of the assets in an industry. This book provides Foreign Investment Ratio information for the United States across industries in Chapter 3.

1.4.7   Imports and Import Ratio

Imports refers to the value of services that have been transferred from a foreign country (across national borders) to a particular country. Imports serves as a proxy of the extent of international competition that takes place in an industry. It is believed that imports bring in increased rivalry in an industry, potentially reducing profits. Several reasons have been offered to explain why this would be the case under the structure–conduct–performance framework (Porter 1980). First, firm motivations tend to vary across countries. Second, foreign firms, in contrast with local firms, have different factor advantages that local firms may not have, allowing them to compete on the basis of low price or through differentiation. Third, imports create more options for customers, resulting in overcapacity and potentially inciting a price war among firms (Elango and Pattnaik 2013). While Imports represents the aggregate value of services brought into a country, it does not offer a relative assessment as to what extent imports dominate a particular industry. Import Ratio presents this information by measuring the extent of market share held by imports in an industry. An industry which has imports controlling 15 % of industry sales will have a score of .15. This book reports Imports for the United States in Chapter 3 and Import Ratio for the United States and OECD nations across industries in Chapter 4.

1.4.8   Industry Size

Industry Size represents the total value of output generated by the industry. It denotes the extent of economic activity that takes place in that industry and typically sets the overall cap for collective firm growth and potential at a point in time, even though industries can grow or shrink in the long run (Elango 1998). This book provides Industry Size information for the United States across industries in Chapter 3.

1.4.9   Labor Ratio

Labor Ratio is measured by dividing compensation of employees by output of the industry. Labor Ratio refers to the cost of labor faced by a firm on its outputs. This ratio illustrates the degree of labor costs in the production of a service (Elango and Abel, 2004). Higher labor costs could be due to the fact that the particular service is very labor intensive or requires high cost (education or skilled) labor to produce services. This variable has also been referred to as human capital intensity (Elango 2005). An industry which has a labor cost representing about 32 % of the value of production will have a Labor Ratio of .32. This book provides Labor Ratio information for the United States across industries in Chapter 3.

1.4.10 Leverage

Leverage represents the extent to which a firm is using borrowed money (Elango et al. 2013). Leverage allows a firm to deploy additional capital in its operations, potentially allowing its investors to earn a higher return. However, this borrowed capital incurs a cost to firms by way of interest payments. Therefore, firms which are highly leveraged face a higher bankruptcy risk. In this book, Leverage is measured as the ratio of total liabilities owned by a firm to total assets held by a firm. In the case where a firm has liabilities equal to 50 % of its assets, its Leverage will be reported as .5. This book provides Leverage information globally across industries in Chapter 5.

1.4.11 Profit Margin Before Tax

Profit Margin Before Tax refers to the profits made before payment of taxes on a firm's sales. Profit Margin indicates the extent to which an industry is profitable, as it represents the portion of revenue left with the firm after all expenses have been paid, before taxes. This variable indicates the extent to which there is managerial slack in the firm (Elango 2000). Stated differently, it is indicative of the extent which a firm can tolerate higher costs or potentially lower sales. In many instances, the variable profit margin is reported on an after-tax basis, but in this book, the before-tax figures are provided, as they are a cleaner measure, not being impacted by varying tax treatments. A firm whose profits before taxes are 10 cents on every dollar of sales will thus have a Profit Margin of 10 %. This value is expressed as a percentage in this book and presented in Chapter 3 for U.S. service industries.

1.4.12 Selling and Administrative Ratio

Selling and Administrative Ratio refers to the extent of expenditures incurred by a firm for selling and administrative purposes per unit of sales. This ratio represents the costs incurred by firms to promote the service and support managerial overhead. It accounts for all expenses a firm has incurred due to activities that are not directly linked to the production of services. The inverse of this ratio has been used as a measure of efficiency score of a firm's selling and administrative ability (Elango 2000).

A firm which spends 10 % of revenue to sustain its managerial overhead and sales activities will have a Selling and Administrative Ratio of .1. This book provides Selling and Administrative Ratio information globally across industries in Chapter 5.

1.5   Industry Classifications

The book uses two industry classifications in the reporting of the data: International Standard Industrial Classification (ISIC) and Standard Industrial Classification (SIC).

The ISIC classification was developed by the United Nations and is currently in its fourth iteration. ISIC industry categories follow a hierarchical four-level structure. At the first level, activities are split into broad groupings (Manufacturing, Agriculture, Financial and insurance activities, etc.). In each subsequent level, they are organized into narrower categories, namely two-digit divisions, three-digit groups, and four-digit classes (highest detail). Example: Section K represents financial and insurance activities (highest level). Within this section there are several two-digit divisions (2nd level), one of which is 65 representing insurance, reinsurance and pension funding, except compulsory social security. This section is further divided into several three-digit groups (3rd level). One of these, 651, represents insurance. More detailed classification is available using several classes in the next (4th) level. In this level, for example, 6511 refers to life insurance and 6512 refers to non-life insurance. Data reported in Chapter 4 of this book follows ISIC Rev. 4 codes at the 2-Digit level. Specific details on the ISIC Rev. 4 classification can be reviewed at http://unstats.un.org/unsd/cr/registry/isic-4.asp.

The SIC classification was initially developed by the United States government and is still widely followed in the United States. The final iteration of SIC was done in 1987. It is organized such that industry groupings can be represented using codes which vary from 2 to 4 digits. At a 2-digit level, the major industry sector is represented. The 3-digit and 4-digit codes represent further sub-categories within an industry. For instance, SIC 60 represents depository institutions (i.e., banks). SIC 60 is then divided into several groups (e.g., SIC 601, SIC 602, SIC 603 and so forth). SIC 603 represents savings institutions (i.e., a particular type of bank), which is further subdivided into two subgroups, SIC 6035 representing savings institutions which are federally chartered and SIC 6036 savings institutions which are not federally chartered. Data reported in Chapter 5 of this book follows SIC codes at the 2-Digit and 3-Digit Level. Specific details on the SIC structure can be reviewed at https://www.osha.gov/pls/imis/sic_manual.html.

1.6   Limitations and Other Disclaimers

Many words of caution are due to readers when using a book of this nature. Any reader should be fully aware of the underlying inherent limitations of a work which reports industry ratios across many countries.

First, it should be noted that the author is not the one collecting the underlying data. The author is essentially compiling data from various sources and developing industry-specific ratios with that data. To illustrate this point further, let us take the case of something as simple as a country's GDP. Such information is not collected by any individual researcher but is accessed from institutions which are tasked with providing such information. Such institutions are typically deemed reliable and the data provided is taken for granted to be valid. Additionally, in many situations there are few options to cross-check this data, as these institutions are the only sources of such information. Even assuming someone is a nation-specific expert on these numbers, only expert judgment or proxy comparisons can be used to check the reliability and reasonableness of the numbers, but they cannot be verified personally. Moreover, it should be noted that in many cases, these numbers are estimates of economic activity and tend to be revised upward or downward as and when more reliable information becomes available by the institutions providing the same.

Second, any data at the industry level is driven by industry classifications. The classifications are not developed by the author. They are standard classifications which were used for the collection and reporting of data by various sources. It is quite possible they are not precise enough to capture every nuance and are subject to some vagaries in interpretation of industry boundaries. As a result, firms could be potentially misclassified into incorrect industry groups. For instance, it is quite possible that two companies who are direct rivals may be classified into different industries. While there is no reason to believe there are systemic errors in the assignment of firms to industries, it should be noted there is no option for the author to ascertain if this assignment was done correctly by the reporting bodies, nor is it reasonable to develop newer industry classifications which are meaningful.

Third, the information provided in Chapter 5 is based on commercial databases which report actual sales figures of firms and therefore can be deemed very reliable. However, two notable issues should be factored in when using this information. Commercial databases do not capture economic activities of private firms, as they, unlike public firms, have no fiduciary duty to report financial data. Additionally, many of the larger public firms listed in the database tend to operate in a multitude of industries. These firms are assigned to their primary industry of operations in computation of the ratios, as it is extremely cumbersome and nearly impossible to differentiate segment sales for many firms.

Fourth, it should be noted that the ratios presented in this book are based on data derived from varied sources and the requisite caveats apply. While it is possible for an individual to develop domain-specific expertise, it is not possible for any one scholar to be knowledgeable about every other service industry across the nations of the world. Such concerns are relevant for a book of this nature, and the author humbly accepts that there is a chance that errors/deviations could have crept into the computation of ratios presented. In instances where the author believes there are systemic errors in the data for a particular year, these values were deleted from the tables presented in the book. However, it should also be noted that reasonable efforts have been made to ensure these errors (if any) have been minimized. First, the underlying data used in the computation of ratios is from the most reputable sources for such data. Second, after computation of the ratios, cross-checking and relative comparisons were done to test the integrity of the data presented. Therefore, while the likelihood of errors is minimal, given the nature of such ratios, the reader may wish to take a careful approach to interpreting the values of various ratios. For example, if the profit ratio for a particular industry is reported as 10.2 % for a particular year, the reader may want to add a range (e.g., ± .5) contingent on factors such as the previous year's numbers and relative numbers from comparable countries when making forecasts or assessments of the reader's own specific context.

Finally, the data presented in the book is limited to the nations where service industry segment-specific data is available. Additionally, in several of the tables, data may be missing for a particular year, contingent on availability of data. While the absence of data on a particular country may be disappointing to those seeking specific figures, it should be noted that, until recently, this was a problem even for the countries which are covered here. Emerging markets, which offer great potential for continued growth of the service sector (Elango and Jones 2011), are not covered in this book due to lack of sector-specific data. Additionally, it is quite possible that many readers may desire more current data than is presented in this book. While this is desirable from the author's point of view, two caveats should be noted. First of all, the data presented in this book is derived data and therefore limited to what is available and second, sector-specific data on a multitude of countries takes time to compile and typically has lag periods greater than 2–3 years.

Despite these limitations, this book will serve as an important reference on service sector industries. The global nature of the data compiled in this book, especially its extensive coverage of the United States, makes it an invaluable resource to active researchers and stakeholders in the service industry as well as those who seek to enter it. In particular, it is a handy reference book for any library catering to individuals involved in strategic planning or performance assessments, financial analysts, consultants, public policymakers or academics working on research within the service industries globally.

References

Attaran M, Saghafi MM (1988) Concentration trends and profitability in the US manufacturing sector. Appl Econ 20(11):1497–1510
Bain JS (1959) Industrial organization. Wiley, New York
Curry B, George KD (1983) Industrial concentration: a survey. J Ind Econ 31(3):203–255
Dalton JA, Penn DW (1976) The concentration-profitability relationship: is there a critical concentration ratio? J Ind Econ 25(2):133–142
Elango B (1998) Influence of industry and firm drivers on the rate of internationalization of U.S. firms. J Int Manag 1(4):201–221
Elango B (2000) An exploratory study into the linkages between corporate resources and the extent and form of internationalization. Am Bus Rev XVIII(2):12–26
Elango B (2005) The influence of plant characteristics on the entry mode choice of manufacturing firms. J Oper Manag 23(1):65–79
Elango B, Abel I (2004) A comparative analysis of the influence of the country characteristics on service versus manufacturing investments. Am Bus Rev 22(2):29–39

Elango B, Jones J (2011) Drivers of insurance demand in emerging markets. J Serv Sci 3(2):185–204
Elango B, Pattnaik C (2013) Response strategies of local firms to import competition in emerging markets. J Bus Res 66(12):2460–2465
Elango B, Sethi SP (2007) Exploration into the relationship between country of origin (COE) and the internationalization-performance paradigm. Manag Int Rev 47(3):369–392
Elango B, Talluri S, Hult TM (2013) Understanding drivers of risk-adjusted performance for service firms with international operations. Decis Sci 44(4):755–783
Porter ME (1980) Competitive strategy. Free Press, New York
Scherer FM (1980) Industrial market structure and economic performance. Rand McNally & Co, Chicago
Shapiro H, Taylor L (1990) The state and industrial strategy. World Dev 18:861–878
United Nations Conference on Trade and Development (UNCTAD). http://unctad.org/
Weiss LW (1979) The structure-conduct-performance paradigm and antitrust. Univ Pa Law Rev 127(4):1104–1140

Data Source

United Nations Conference on Trade and Development (UNCTAD). http://unctad.org/en/Pages/Statistics.aspx

Potential Sources of Data on the Service Sector

Data sources for the service sector are presented in two groups, Macro and Micro, based on the level (i.e., firm, industry, country) of data sought. Therefore, when targeting a particular source, the reader may first want to ask what type/level of data they are looking for. If looking for aggregate information (country-, industry-, or region-specific information), the reader may want to focus on the Macro sources listed. If looking for firm-specific information, the reader may want to focus on the Micro sources listed. Figure 2.1 provides a flow chart of this process.

2.1 Macro Sources

2.1.1 Transnational Organizations

As a group, transnational organizations offer service sector data at the national level. These sources are typically the only sources for such data but are usually deemed very reliable. However, in a few rare instances, the macro sources may offer firm-specific information. In such cases, the reader may be presented with a table with firm-specific information. Even when such a table exists, however, they tend to be selective in their choice of firms and usually limited in coverage. As noted in Chapter 1, readers should be aware that the data presented in these sources have a time lag of about 2–4 years, due to the time it takes to consolidate information across countries. These sources include the following:

2.1.1.1 World Bank
The World Bank (http://data.worldbank.org/) provides information on the service sector for the following variables: exports/imports of services (% of GDP); value added by services (% of GDP); and trade in services (% of GDP). Coverage is for most of the countries of the world and many time periods, allowing for time series analysis when needed. This data is free for users.

© Springer International Publishing Switzerland 2015   13
B. Elango, *Service Industry Databook: Understanding and Analyzing Sector Specific Data Across 15 Nations*, DOI 10.1007/978-3-319-19111-9_2

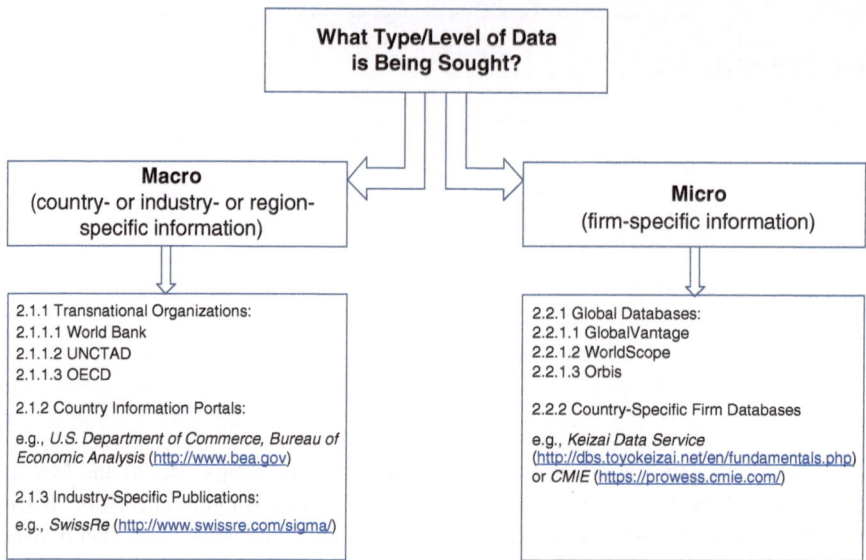

Fig. 2.1 Data sources flow chart

2.1.1.2 UNCTAD

United Nations Conference on Trade and Development (http://unctad.org) regularly publishes reports on specific service sectors and holds several related conferences. While the degree of interest of specific activities of UNCTAD may vary from reader to reader, some events may contain useful nuggets of information on the service sector and should not be ignored. Importantly, UNCTAD has a specific section on its website covering data on trade in services. Here data is presented on service sectors using broad generic categories (creative services, computer information, royalties, etc.) and is offered across countries and time periods. This data is free for users.

2.1.1.3 OECD

Organisation for Economic Co-operation and Development (http://www.oecd.org) offers useful sector-specific data on the service sector. In fact, to the knowledge of the author, it is the only source offering time series data on several countries by service industry segment. The data presented in Chapter 4 is derived from OECD statistics. It should be noted that while much data offered by OECD can be accessed for free, the sector-specific data used in the development of certain ratios in this book is not and requires a fee.

2.1.2 Country Information Portals

Most countries offer an information portal which presents statistical information on that country. These sites are broad in terms of information presented, and thus researchers will need to comb through the information presented to find data on specific service sectors or services. However, one exceptional site with data on the United States service sector is the U.S. Department of Commerce, Bureau of Economic Analysis (http://www.bea.gov). Data presented in Chapter 3 is derived from information from the Bureau of Economic Analysis. Other examples of country-specific data statistics include: http://www.statistics.gov.uk for information on the UK, http://www.insee.fr/en/ for France, http://eng.stat.gov.tw for Taiwan, http://kostat.go.kr/portal/english/index. action for Korea, http://www.stat.go.jp/english/ for Japan, etc. Readers are warned that web links reported here were accurate at the time of writing, but may no longer work after publication, as most websites frequently change their specific webpage links, often rendering the presentation of such links in a published book worthless. In such instances, readers may need to locate the website using an internet search engine.

Friendly advice for readers going to these websites looking for data is to have modest expectations and be willing to use proxies to capture the information of interest. For instance, a reader may be in interested in the size of a particular service segment's sales. The reader may only be able to locate the number of people employed in this sector and therefore should be willing to estimate sales figures based on people employed. While such a figure cannot be totally accurate, the question then is, without other easy options, would such an estimate suffice? This is clearly a call to be made by the reader.

2.1.3   Industry-Specific Publications

Another interesting avenue for data on the service sector is industry-oriented portals. While such websites might not be available for every industry in the service sector and the information presented is typically limited to one industry, the information presented in such sites may make it worthwhile for researchers. For example, the Switzerland-based global insurance company SwissRe publishes a newsletter entitled Sigma (http://www.swissre.com/sigma/). Sigma, in one of its issues, annually publishes detailed information on insurance sales by insurance segment across many countries of the world.

2.2   Micro Sources

2.2.1   Global Databases

Firm-level databases usually offer the cleanest source of data available for research, as they are based on financial numbers reported by firms, which are required to follow particular accounting standards. With these databases, a researcher is able to visually see the data firm- and time-wise, allowing the identification and removal of anomalies. These databases offer search tools allowing researchers to identify, select and download data on service firms for analysis. Some of these packages allow for custom-interfacing with spreadsheet packages (e.g., Excel) for direct data download. It is also quite possible to use firm-level information and gather industry- or country-level data, with reasonable effort. Chapter 5 uses this approach. However, access to these databases is not free and typically requires a subscription payment of several thousands of dollars.

2.2.1.1 GlobalVantage

S&P Capital IQ, a division of McGraw-Hill, offers financial data on firms world-wide covering several years in its GlobalVantage database (https://www.capitaliq.com/). GlobalVantage claims to include about 41,100 active public companies globally, which covers about 99 % of world market capitalization.

2.2.1.2 WorldScope

Thomson Reuters (http://www.trpropresearch.com/products/Worldscope/) provides financial data on both public and private companies globally through its WorldScope database. WorldScope claims to contain up to 20 years of data on 57,000 companies, representing about 95 % of market value.

2.2.1.3 Orbis

Bureau Van Dijk's Orbis database provides financial data on firms globally (http://www.bvdinfo.com/en-us/our-products/company-information). Orbis claims to provide detailed information on 65,000 listed companies and selective information on millions of private firms worldwide.

2.2.2   Country-Specific Firm Databases

While the above three options cover companies globally, researchers should not ignore country-specific options that may be available on a country-by-country basis. In many cases, country-specific options may cover a greater number of firms in a particular nation compared to global databases. For instance, in the case of Japanese firms, financial information is available from Toyo Keizai Data Service (http://dbs.toyokeizai.net/en/fundamentals.php) and in the case of Indian firms, financial information is available from CMIE's Prowess database (https://prowess.cmie.com/) and Capital Market Publishers' Capitaline database (http://www.capitaline.com/new/index.asp).

United States Service Sector Data

3.1   Overview

This chapter provides sector-specific data for the United States. The data is either derived or reported directly from U.S. Government sources. Most of the data presented in this chapter comes from the U.S. Department of Commerce. The data on the United States is more detailed than that for any other country provided in this book. It should be noted that the United States data located in Chapters 3 and 4 may not be comparable, as Chapter 3 uses industry classifications followed by U.S. Government Sources, while Chapter 4 is derived from OECD sources and follows ISIC (Rev. 4) classifications.

3.2   Exports

Service exports from the United States are presented in Table 3.1a for the years 1999–2005 and in Table 3.1b for the years 2006–2013, in millions of US dollars. All figures reported in the tables are collected from the Bureau of Economic Analysis, U.S. Department Commerce. U.S. service exports were at $271,343 million in 1999, which increased to $687,410 million in 2013. While U.S. exports more than doubled in the period for which data is provided, the year-to-year growth was not always positive. Figure 3.1 illustrates graphically the overall growth of exports, and Figure 3.2 graphically illustrates the year-to-year changes in export growth rate.

© Springer International Publishing Switzerland 2015
B. Elango, *Service Industry Databook: Understanding and Analyzing Sector Specific Data Across 15 Nations*, DOI 10.1007/978-3-319-19111-9_3

Table 3.1a U.S. service exports (1999–2005) [Millions of U.S. dollars]

Industry	1999	2000	2001	2002	2003	2004	2005
Maintenance and repair services	3812	4686	5575	5769	5458	5342	7218
Transport	43,218	45,758	41,716	41,912	41,446	47,723	52,622
Sea transport	10,489	11,554	10,946	11,187	10,900	11,801	12,207
Sea transport: Freight	3951	4301	3780	3733	4019	3800	3463
Sea transport: Port	6538	7253	7166	7454	6881	8001	8744
Air transport	29,786	31,267	27,752	27,560	27,379	32,269	36,632
Air transport: Passenger	19,425	20,197	17,181	16,291	15,091	17,932	20,609
Air transport: Freight	5044	5776	5428	5787	6869	7726	9327
Air transport: Port	5317	5294	5143	5482	5419	6611	6696
Other modes of transport	2942	2938	3017	3164	3166	3653	3783
Travel (for all purposes including education)	92,338	100,187	86,733	81,869	80,332	92,387	101,470
Business	39,089	42,664	35,784	32,572	30,379	33,949	3,8151
Expenditures by border, seasonal, and other short-term workers	5921	5975	6329	6694	6762	7363	8435
Other business travel	33,168	36,689	29,455	25,878	23,617	26,586	29,716
Personal	53,249	57,523	50,949	49,297	49,954	58,437	63,320
Health-related	1351	1501	1479	1460	1571	1689	2014
Education-related	9616	10,348	11,476	12,626	13,312	13,633	14,022
Other personal travel	42,282	45,674	37,994	35,211	35,071	43,115	47,284
Insurance services	3052	3631	3424	4415	5974	7314	7566
Direct insurance	436	628	570	877	1152	1760	2696
Reinsurance	2616	3002	2731	3339	4383	4742	4276
Auxiliary insurance services			122	200	439	812	594
Financial services	19,433	22,117	21,899	24,496	27,840	36,389	39,878

Industry	1999	2000	2001	2002	2003	2004	2005
Charges for the use of intellectual property	47,731	51,808	49,489	53,859	56,813	67,094	74,448
Telecommunications, computer, and information services	12,287	12,215	12,829	12,451	14,061	14,962	15,515
Telecommunications services	5599	5266	5982	5372	5848	6269	6081
Computer services	4340	4358	4197	4380	4953	5100	5301
Information services	2348	2591	2650	2699	3260	3593	4133
Other business services	40,976	40,497	44,146	47,996	48,775	54,398	58,302
Research and development services	6563	7116	7610	8678	9467	9563	10,431
Professional and management consulting services	15,689	17,621	18,484	19,874	20,240	23,397	27,103
Technical, trade-related, and other business services	18,723	15,760	18,052	19,444	19,068	21,437	20,768
Government goods and services	8495	9481	8514	7903	9274	12,357	15,989
Exports of services	271,343	290,381	274,323	280,670	289,972	337,966	373,006

Data Source: U.S. Department of Commerce

Table 3.1b U.S. service exports (2006–2013) [Millions of U.S. dollars]

Industry	2006	2007	2008	2009	2010	2011	2012	2013
Maintenance and repair services	7673	9062	10,019	12,077	13,860	14,279	15,115	16,295
Transport	57,462	65,824	74,973	62,189	71,656	79,830	83,592	87,267
Sea transport	14,616	16,412	17,779	13,603	15,905	16,460	17,055	17,175
Sea transport: Freight	3407	4213	4867	3249	4229	4124	4178	3987
Sea transport: Port	11,209	12,199	12,912	10,354	11,676	12,336	12,877	13,188
Air transport	38,855	45,330	52,842	44,574	51,579	59,015	61,683	65,522
Air transport: Passenger	21,638	25,187	30,957	26,103	30,987	36,763	39,364	41,642
Air transport: Freight	10,076	11,311	13,483	10,941	12,280	13,482	13,871	14,321
Air transport: Port	7141	8832	8402	7530	8312	8770	8448	9559
Other modes of transport	3991	4083	4352	4013	4172	4355	4854	4570
Travel (for all purposes including education)	105,140	119,037	133,761	119,902	137,010	150,867	161,249	173,131
Business	39,333	42,819	45,941	36,518	39,523	40,281	39,272	39,396
Expenditures by border, seasonal, and other short-term workers	8700	7984	8655	6843	6344	6367	6633	7018
Other business travel	30,633	34,835	37,286	29,675	33,179	33,914	32,639	32,378
Personal	65,807	76,218	87,820	83,384	97,487	110,586	121,978	133,736
Health-related	2221	2365	2531	2691	2876	3032	3176	3312
Education-related	14,647	15,956	17,956	19,689	20,937	22,823	24,858	27,241
Other personal travel	48,939	57,897	67,334	61,004	73,674	84,731	93,944	103,182
Insurance services	9445	10,841	13,403	14,586	14,397	15,114	16,534	16,096
Direct insurance	3087	3224	3792	4069	4029	4129	4038	4203
Reinsurance	5461	6678	8087	8895	8981	9345	11,077	10,505
Auxiliary insurance services	897	939	1524	1622	1387	1640	1420	1388
Financial services	47,882	61,376	63,027	64,437	72,348	78,271	76,605	84,066
Securities brokerage, underwriting, and related services	15,554	19,037	19,269	18,927	20,885	21,224	16,066	15,879
Financial management, financial advisory, and custody services	19,431	26,574	25,030	23,582	27,319	28,492	28,971	33,682

Credit card and other credit-related services	4928	5749	7660	9402	10,460	13,506	15,542	16,864
Securities lending, electronic funds transfer, and other services	7968	10,016	11,068	12,527	13,684	15,050	16,026	17,640
Charges for the use of intellectual property	83,549	97,803	102,125	98,406	107,521	123,333	125,492	129,178
Industrial processes	32,415	36,160	37,685	34,865	36,333	43,897	43,252	44,978
Computer software	22,655	29,466	31,414	31,017	36,008	39,459	40,506	42,907
Trademarks and franchise fees	13,653	15,591	17,652	16,657	19,131	21,900	22,738	22,799
Audio-visual and related products	14,720	16,498	15,288	15,746	15,886	17,865	18,884	18,398
Other intellectual property	106	88	86	120	163	212	113	97
Telecommunications, computer, and information services	17,184	20,192	23,119	23,816	25,038	29,171	32,103	33,409
Telecommunications services	7105	8239	9999	10,102	10,911	12,424	13,756	14,154
Computer services	5734	7203	8502	8821	8991	11,361	12,086	12,346
Information services	4344	4750	4618	4893	5137	5386	6261	6909
Other business services	68,619	82,382	92,738	95,984	101,029	112,568	119,892	123,447
Research and development services	12,810	15,625	17,345	18,636	22,446	25,761	28,080	30,052
Professional and management consulting services	31,167	38,900	42,684	42,837	46,749	49,645	53,539	55,758
Technical, trade-related, and other business services	24,642	27,857	32,709	34,511	31,834	37,162	38,273	37,637
Government goods and services	19,783	21,879	19,652	21,324	20,474	24,348	24,267	24,522
Exports of services	416,738	488,396	532,817	512,722	563,333	627,781	654,850	687,410

Data Source: U.S. Department of Commerce

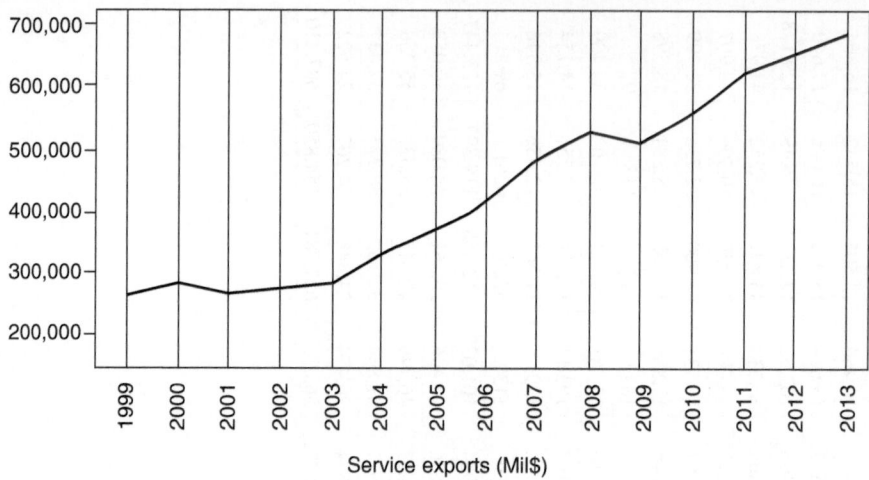

Fig. 3.1   U.S. service exports for the years 1999–2013

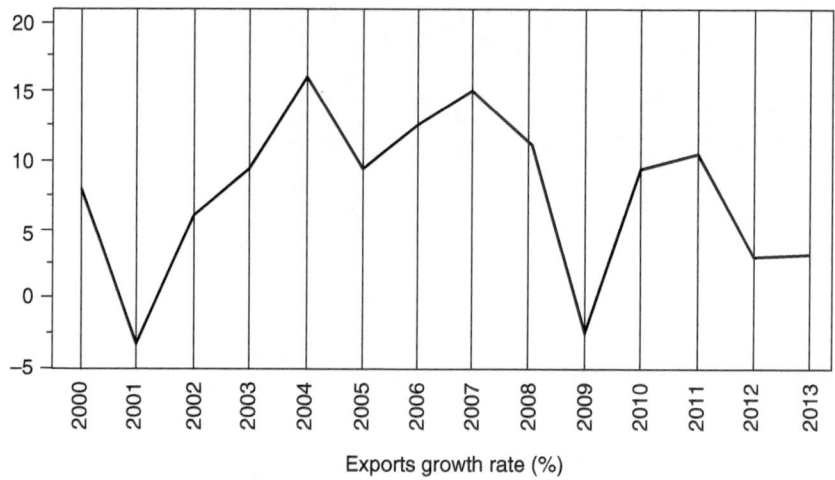

Fig. 3.2   U.S. service exports growth rate for the years 2000–2013

3.3   Imports

Service imports from the United States are presented in Table 3.2a for the years 1999–2005 and in Table 3.2b for the years 2006–2013, in millions of US dollars. All figures reported in the tables are collected from the Bureau of Economic Analysis, U.S. Department Commerce. U.S. service imports were at $192,893 million in 1999, which increased to $462,134 million in 2013. Similar to U.S. exports, imports more than doubled in the period for which data is provided, with much variation in year-to-year growth. Figure 3.3 graphically illustrates the overall growth of imports, and Figure 3.4 graphically illustrates the year-to-year changes in import growth rate.

Table 3.2a U.S. service imports (1999–2005) [Millions of U.S. dollars]

Industry	1999	2000	2001	2002	2003	2004	2005
Maintenance and repair services	1278	2569	1999	2217	2246	2395	3015
Transport	49,620	57,606	53,840	51,491	57,863	69,158	75,643
Sea transport	17,405	21,787	20,824	20,100	25,566	31,710	35,687
Sea transport: Freight	15,760	20,100	19,425	18,651	24,193	30,373	34,311
Sea transport: Port	1645	1687	1399	1449	1373	1337	1376
Air transport	29,350	32,678	30,136	28,481	29,199	34,045	36,103
Air transport: Passenger	18,126	20,397	18,931	16,439	17,244	20,213	21,431
Air transport: Freight	4138	4738	3959	4880	5018	5978	6113
Air transport: Port	7086	7543	7246	7162	6937	7854	8559
Other modes of transport	2865	3141	2880	2910	3098	3402	3853
Travel (for all purposes including education)	59,592	65,787	60,730	59,942	61,884	74,024	79,988
Business	21,216	22,640	20,031	18,681	18,476	21,811	22,813
Expenditures by border, seasonal, and other short-term workers	342	366	389	323	502	584	627
Other business travel	20,874	22,274	19,642	18,358	17,974	21,227	22,186
Personal	38,377	43,147	40,699	41,260	43,408	52,213	57,175
Health-related	140	155	156	152	168	421	510
Education-related	1808	2032	2300	2702	3148	3542	3992
Other personal travel	36,429	40,960	38,243	38,406	40,092	48,250	52,673
Insurance services	9389	11,284	16,706	21,927	25,233	29,089	28,710
Direct insurance	1452	1683	2007	3771	3634	3162	2967
Reinsurance	7937	9601	14,523	17,798	21,070	25,280	25,318
Auxiliary insurance services			176	357	529	647	425
Financial services	8280	10,936	10,157	8963	8948	11,156	12,126
Charges for the use of intellectual property	13,302	16,606	16,661	19,493	19,259	23,691	25,577

(continued)

Table 3.2a (continued)

Industry	1999	2000	2001	2002	2003	2004	2005
Telecommunications, computer, and information services	13,332	12,397	12,421	11,721	13,063	14,210	15,975
Telecommunications services	7290	6167	5910	5226	5447	5572	5380
Computer services	5750	5934	6134	6149	7259	8180	10,028
Information services	291	295	376	346	357	458	567
Other business services	23,887	24,414	25,629	29,274	30,103	33,065	35,960
Research and development services	3274	3346	3389	4063	5071	5778	7239
Professional and management consulting services	10,539	10,648	10,986	13,735	13,740	15,419	18,421
Technical, trade-related, and other business services	10,074	10,420	11,254	11,475	11,292	11,869	10,300
Government goods and services	14,212	14,516	15,322	19,353	23,619	26,296	27,454
Imports of services	192,893	216,115	213,465	224,379	242,219	283,083	304,448

Data Source: U.S. Department of Commerce

Table 3.2b U.S. service imports (2006–2013) [Millions of U.S. dollars]

Industry	2006	2007	2008	2009	2010	2011	2012	2013
Maintenance and repair services	4583	5209	5742	5938	6909	8236	7970	7620
Transport	77,962	79,326	83,988	64,133	74,628	81,377	85,029	90,754
Sea transport	35,996	34,760	34,463	23,219	29,496	31,369	33,206	36,256
Sea transport: Freight	34,274	32,888	32,496	21,656	27,872	29,703	31,401	34,189
Sea transport: Port	1723	1872	1967	1563	1624	1666	1805	2067
Air transport	37,929	40,520	45,493	37,410	41,377	45,834	47,458	50,232
Air transport: Passenger	22,642	23,553	27,292	21,532	23,426	26,747	29,565	32,029
Air transport: Freight	6266	6376	6179	4688	6435	6613	6177	6325
Air transport: Port	9021	10,591	12,022	11,190	11,516	12,474	11,716	11,878
Other modes of transport	4036	4047	4031	3503	3755	4174	4365	4266
Travel (for all purposes including education)	84,206	89,235	92,545	81,421	86,623	89,700	100,317	104,677
Business	23,047	23,999	23,204	19,273	21,274	20,675	20,104	21,228
Expenditures by border, seasonal, and other short-term workers	819	909	981	1013	1065	1116	1148	1195
Other business travel	22,228	23,090	22,223	18,260	20,209	19,559	18,956	20,033
Personal	61,158	65,236	69,341	62,148	65,349	69,024	80,213	83,449
Health-related	585	660	757	879	1019	1139	1282	1443
Education-related	4467	4725	5143	5144	5468	5782	6103	6490
Other personal travel	56,107	59,851	63,441	56,125	58,862	62,103	72,828	75,516
Insurance services	39,382	47,517	58,913	63,801	61,478	55,654	53,203	50,454
Direct insurance	4309	4155	3707	3449	4374	4881	5028	4911
Reinsurance	34,557	42,843	53,725	59,005	55,660	49,157	46,317	44,177
Auxiliary insurance services	517	518	1481	1346	1444	1616	1858	1366
Financial services	14,733	19,197	17,218	14,415	15,502	17,368	16,975	18,683
Securities brokerage, underwriting, and related services	2882	3825	4567	4287	3913	3801	3513	3637

(continued)

Table 3.2b (continued)

Industry	2006	2007	2008	2009	2010	2011	2012	2013
Financial management, financial advisory, and custody services	5922	7094	5236	4165	4105	4375	4478	5354
Credit card and other credit-related services	785	827	817	2068	3702	5245	5332	6160
Securities lending, electronic funds transfer, and other services	5144	7452	6599	3895	3782	3946	3652	3531
Charges for the use of intellectual property	25,038	26,479	29,623	31,297	32,551	36,087	39,502	39,015
Industrial processes	16,535	16,660	16,223	17,914	18,847	21,506	22,262	22,353
Computer software	2973	4799	6104	6203	5228	5697	6888	6633
Trademarks and franchise fees	2132	2424	3585	4137	4635	4782	4495	4643
Audio-visual and related products	3285	2492	3556	2946	3627	3644	5474	5275
Other intellectual property	114	104	155	98	214	458	383	112
Telecommunications, computer, and information services	19,776	22,384	24,655	25,784	29,015	32,756	32,156	32,877
Telecommunications services	6342	7272	7761	7579	7986	7039	7182	7298
Computer services	12,847	14,323	15,925	16,844	19,407	23,879	23,221	23,643
Information services	587	788	970	1361	1622	1839	1753	1936
Other business services	48,130	54,968	67,488	68,553	70,646	83,289	87,347	92,710
Research and development services	9276	13,032	17,122	18,241	22,170	26,558	28,713	32,142
Professional and management consulting services	23,192	24,945	28,926	26,655	27,690	32,270	33,269	34,480
Technical, trade-related, and other business services	15,662	16,991	21,441	23,657	20,786	24,461	25,365	26,088
Government goods and services	27,353	28,260	28,880	31,460	31,960	31,293	27,861	25,343
Imports of services	341,165	372,575	409,052	386,801	409,313	435,761	450,360	462,134

Data Source: U.S. Department of Commerce

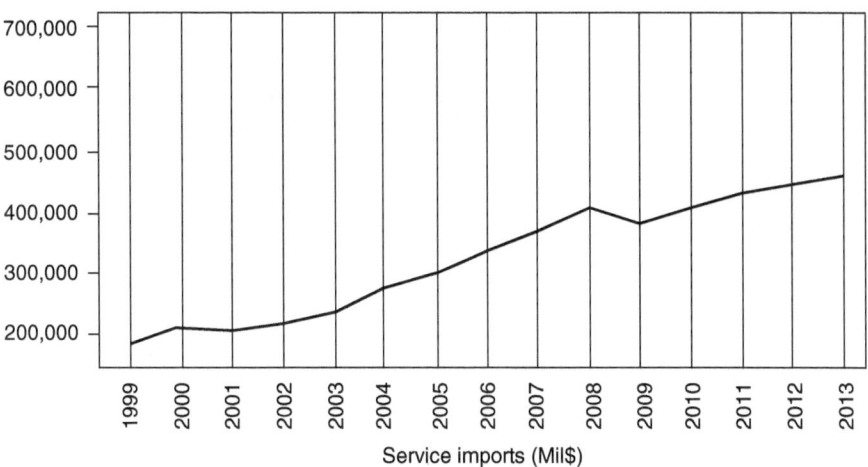

Fig. 3.3 U.S. service imports for the years 1999–2013

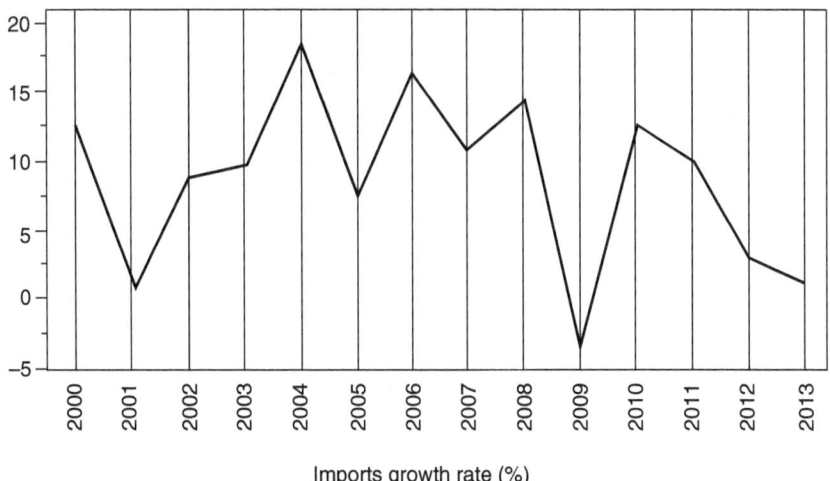

Fig. 3.4 U.S. service imports growth rate for the years 2000–2013

3.4   Export Ratio

Export Ratio represents the ratio of value of services that was transferred outside a country's borders relative to the level of economic output that takes place in a particular industry in a country. Stated differently, it represents the extent of exports that take place in an industry. This variable is calculated by the author as follows: Initially, information on exports and gross productive output of varied sectors are collected from U.S. Department of Commerce sources for the years for which data is available. Then, Export Ratio is computed by dividing exports by gross productive output in the corresponding service sector. Export Ratio for the United States is presented in Table 3.3 for years 2004–2011 [readers are advised that the U.S.

Table 3.3 U.S. export ratio

Industry	2004	2005	2006	2007	2008	2009	2010	2011
Construction			0.056	0.058	0.095	0.103	0.078	0.133
Telecommunications	0.859	0.832	1.199	1.332	1.570	1.613	1.721	1.887
Financial services	1.105	1.090	1.233	1.484	1.569	1.628	1.737	1.827
Insurance services	1.304	1.260	1.515	1.640	2.020	2.344	2.254	2.328
Business and professional services	5.889	6.002	6.332	7.017	7.429	7.929	8.273	8.345
Legal services			2.065	2.395	2.682	2.733	2.669	2.774
Computer/Information systems design and related services			1.984	2.209	2.371	2.696	2.609	2.752
Management, scientific, and technical services	3.410	3.317	3.979	4.302	4.679	5.029	5.340	5.374
By U.S. parents to/from their foreign affiliates	0.530	0.547	0.581	0.689	0.705	0.722	0.737	0.798
By U.S. affiliates to/from their foreign parent groups	2.309	2.138	2.289	2.404	2.655	3.035	3.024	2.975
Services	2.575	2.613	2.735	3.038	3.262	3.230	3.382	3.546

Note: Export Ratio is computed by dividing the exports by the economic activity that takes place in that industry sector and is expressed as a percentage. Export and economic activity data are sourced from U.S. Department of Commerce

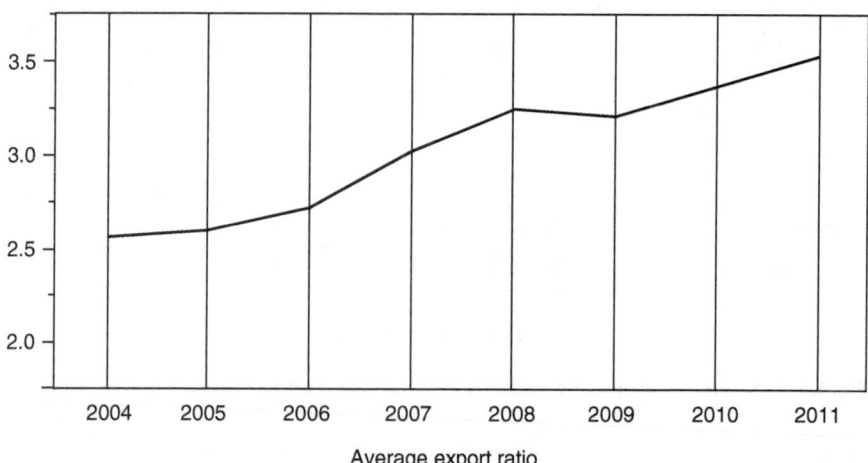

Fig. 3.5 U.S. export ratio for the years 2004–2011

information on this variable is also presented in Chapter 4 (Table 4.15) using OECD data sources]. Figure 3.5 graphically illustrates the Export Ratio across service industries for the years 2004–2011. Export Ratio for the United States varied between 2.575 and 3.546 for the years 2004–2011, and has increased year to year with a marginal reduction in 2009.

3.5 Import Ratio

Import Ratio represents the ratio of the value of services that have been transferred from a foreign country relative to the level of economic output that takes place in a particular industry in a country. This variable captures the import market share in an industry and is computed by the author as follows: Initially, information on imports and gross productive output of varied sectors are gleaned from U.S. Department of Commerce sources for the years for which data is available. Import Ratio is then computed by dividing imports by gross productive output in the corresponding service sector. Import Ratio for the United States is presented in Table 3.4 for years 2004–2011 [readers are advised that the U.S. information on this variable is also presented in Chapter 4 (Table 4.30) using OECD sources]. Figure 3.6 graphically presents Import Ratio across service industries for the years 2004–2011. Import Ratio for the United States varied between 1.969 and 2.375 for the years 2004–2011 and shows an overall positive trend with minor declines or increases on a year-to-year basis.

Table 3.4 U.S. import ratio

Industry	2004	2005	2006	2007	2008	2009	2010	2011
Construction			0.040	0.043	0.060	0.061	0.052	0.064
Telecommunications	0.849	0.792	1.071	1.176	1.219	1.210	1.247	1.147
Financial services	1.018	0.939	1.164	1.371	1.563	1.612	1.551	1.486
Insurance services	5.187	4.781	6.316	7.190	8.880	10.254	9.467	8.518
Business and professional services	3.530	3.748	4.519	4.726	5.290	5.555	5.859	6.505
Legal services			0.481	0.575	0.703	0.617	0.571	0.664
Computer/Information systems design and related services			2.644	2.785	3.054	3.579	3.936	4.356
Management, scientific, and technical services	1.824	1.634	2.147	2.226	2.431	2.607	2.685	2.975
By U.S. parents to/from their foreign affiliates	0.178	0.185	0.247	0.275	0.297	0.322	0.330	0.388
By U.S. affiliates to/from their foreign parent groups	3.004	3.367	3.098	3.079	3.426	3.703	3.786	3.920
Services	1.969	1.963	2.080	2.166	2.330	2.282	2.315	2.375

Note: Import Ratio is computed by dividing the imports by the economic activity that takes place in that industry sector and is expressed as a percentage. Import and economic activity data are sourced from U.S. Department of Commerce

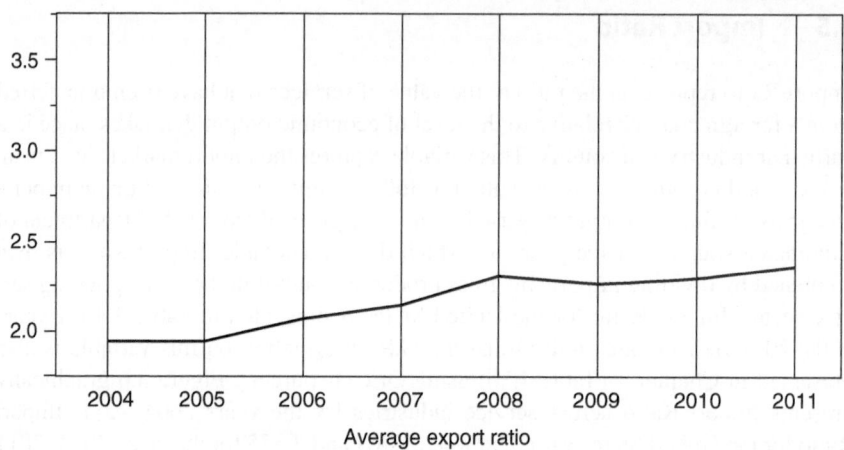

Fig. 3.6 U.S. import ratio for the years 2004–2011

3.6   Foreign Investment Ratio

Foreign Investment Ratio represents the extent of investment made by foreign (non-domiciled) firms in a particular service sector. As many services can be inventoried, this variable serves as a good representation of the level of foreign competition in service industries. This variable is computed by the author as follows: Initially, information on foreign investment position and net stock of private fixed assets of each of the service sector is collected from U.S. Department of Commerce sources for the years for which data is available. This ratio is then computed by dividing foreign investment position in the United States by net stock of private fixed assets in a service industry sector and is expressed as a percentage. Table 3.5a presents U.S. Foreign Investment Ratio for the years 1997–2005 and Table 3.5b presents the same information for years 2006–2013. Figure 3.7 illustrates the foreign investment ratio across service industries graphically for the years 1997–2013. For the years for which data is presented, foreign investment ratio has an average of about 10 with an upward trend in recent years. It was the highest (13.01) in 2000 and lowest in 1997 (7.13).

Table 3.5a U.S. foreign investment ratio (1997–2005)

Industry	1997	1998	1999	2000	2001	2002	2003	2004	2005
Wholesale trade	33.46	35.46	38.18	59.26	62.10	63.26	60.88	68.05	67.89
Information	5.22	5.34	8.29	14.00	13.05	11.76	11.42	11.90	8.16
Publishing industries, except internet (includes software)	34.57	38.93	37.06	37.80	22.98	36.50	26.84	32.68	33.22
Motion picture and sound recording industries	3.77	2.79	10.35	11.42	19.94	14.20	13.54	12.98	1.45
Real estate and rental and leasing	0.84	0.81	0.88	0.85	0.71	0.64	0.52	0.48	0.45
Real estate	0.82	0.80	0.75	0.73	0.59	0.53	0.42	0.39	0.36
Professional, scientific, and technical services	4.31	4.34	4.96	11.36	10.84	11.17	11.60	12.70	13.76
Computer systems design and related services	10.13	9.33	8.93	26.18	13.05	9.33	11.99	11.26	16.01
Legal services	0.10	0.06	0.08		0.07				
Health care and social assistance	1.38	1.42	1.31	1.12	0.97	1.05	0.95	0.88	0.80
Ambulatory health care services		3.80							
Hospitals									
Nursing and residential care facilities			1.32	0.83	0.59	0.58	0.56	0.89	1.05
Social assistance	0.01	0.01	0.01	0.01	0.01	0.01	0.01	0.01	0.01
Accommodation and food services	5.98	5.77	5.23	9.69	9.96	8.58	9.00	6.94	6.85
Accommodation	10.40	10.30	8.68	16.70	15.08	11.38	11.44	8.60	8.62
Food services and drinking places	1.79	1.41	1.77	2.52	4.51	5.47	6.24	5.01	4.76
Educational services	0.07				0.05	0.02	0.03	0.06	0.14
Arts, entertainment, and recreation	2.85					1.19	1.16	1.14	1.05
Amusements, gambling, and recreation industries	5.59	3.41	2.64	2.77	2.01	1.47	1.38	1.39	1.09

Note: Foreign Investment Ratio is computed by dividing foreign investment position in the United States by net stock of private fixed assets in that service industry sector and is expressed as a percentage. Foreign investment position and net stock of private fixed assets data are sourced from U.S. Department of Commerce

Table 3.5b   U.S. foreign investment ratio (2006–2013)

Industry	2006	2007	2008	2009	2010	2011	2012	2013
Wholesale trade	69.93	80.46	81.59	62.24	66.23	70.08	70.21	72.95
Information	10.35	10.81	11.08	9.45	8.37	8.30	8.30	9.21
Publishing industries, except internet (includes software)	39.63	34.86	37.10	27.16	18.64	18.58	23.39	29.22
Motion picture and sound recording industries	1.50				5.94			
Real estate and rental and leasing	0.47	0.60	0.50	0.46	0.45	0.42	0.46	0.48
Real estate	0.40	0.44	0.38	0.36	0.37	0.34	0.36	0.36
Professional, scientific, and technical services	12.03	12.73	13.04	16.24	19.36	17.64	20.30	20.98 .
Computer systems design and related services	19.02	25.19	23.41	29.48	53.35	37.40	52.91	55.83
Legal services		0.28	0.38	0.43	0.46			
Health care and social assistance	0.34			0.88	0.64	0.66	1.01	1.34
Ambulatory health care services	2.53				2.23	2.18	3.70	5.04
Hospitals		0.06						0.04
Nursing and residential care facilities		0.18				1.41		0.55
Social assistance	0.01	0.02	0.02	0.02	0.02	0.02	0.02	0.02
Accommodation and food services	6.30	4.88	4.17	3.68	3.97	4.78	4.33	4.57
Accommodation	7.48	4.82	3.53	3.17	2.97	2.99	2.33	2.56
Food services and drinking places	4.79	4.96	5.15	4.49	5.56	7.56	7.31	7.45
Educational services	0.25		1.54	1.45	1.51	0.93	0.77	0.74
Arts, entertainment, and recreation	0.97	1.00	0.72	0.54	0.19	1.22	1.25	1.23
Amusements, gambling, and recreation industries	1.10	1.52	1.00		0.28	2.37	2.45	2.40

Note: Foreign Investment Ratio is computed by dividing foreign investment position in the United States by net stock of private fixed assets in that service industry sector and is expressed as a percentage. Foreign investment position and net stock of private fixed assets data are sourced from U.S. Department of Commerce

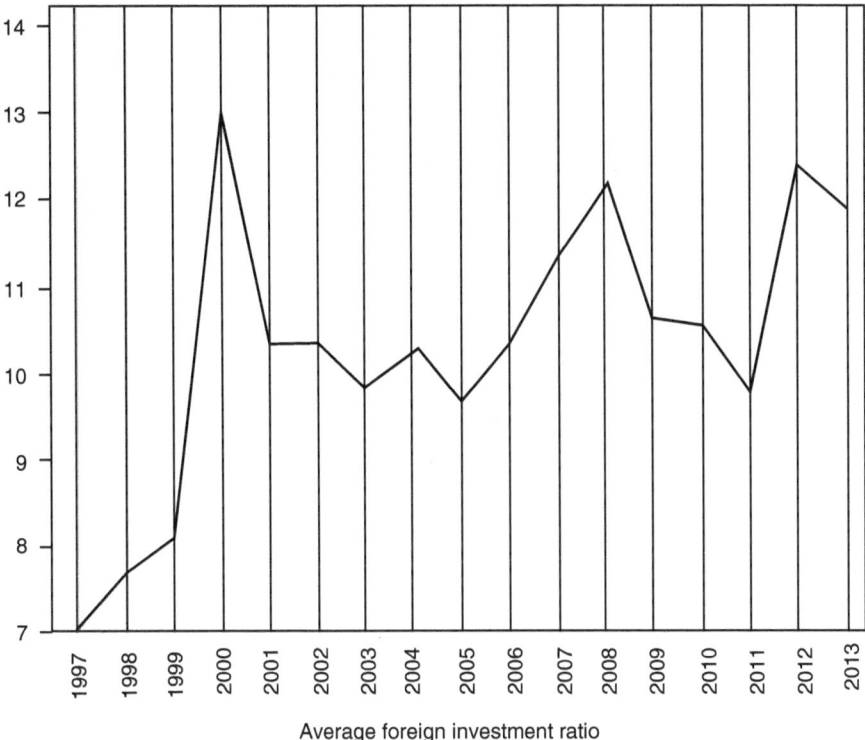

Average foreign investment ratio

Fig. 3.7   U.S. foreign investment ratio for the years 1997–2013

3.7   Asset Intensity

Asset Intensity indicates the level of assets needed to generate one unit of sales. This variable, when compared across industries, allows for the comparison of the extent of capital needed to offer services. Industries which have a lower Asset Intensity are less expensive to enter, because they require lower investments in assets relative to the sales they can generate. This variable was calculated by the author as follows: Initially, information on historical cost of private fixed assets and gross output in each sector of the service industry was secured from U.S. Department of Commerce sources. Asset Intensity was then computed by dividing historical cost of private fixed assets by the gross output in that service industry sector and is expressed as a percentage. Table 3.6a presents information on Asset Intensity for the United States for the years 1997–2005 and Table 3.6b presents the same for years 2006–2013. Figure 3.8 presents Asset Intensity across service industries for the years 1997–2012 for service sector firms. Asset Intensity varies from about 75 (1997) to 87 (2009), with a mean of 79. It is quite possible that values for 2009 were inflated due to the economic shock of 2009 and caution should be exercised with this number.

Table 3.6a U.S. asset intensity (1997–2005)

Industry	1997	1998	1999	2000	2001	2002	2003	2004	2005
Wholesale trade	33.51	34.45	33.77	33.20	34.49	33.56	33.33	31.38	31.06
Retail trade	42.16	43.61	43.90	45.63	49.06	49.47	49.70	49.44	50.21
Transportation and warehousing	77.65	80.61	81.50	81.38	86.63	86.88	82.66	76.21	72.06
Air transportation	93.18	110.98	118.32	121.50	153.54	162.76	145.86	131.37	116.17
Railroad transportation	222.31	233.68	238.48	239.65	242.01	244.55	245.86	230.47	211.60
Water transportation	102.85	108.47	105.73	99.64	96.39	98.15	84.35	71.85	68.70
Truck transportation	32.83	32.80	32.64	32.24	32.56	32.52	32.09	31.11	32.92
Transit and ground passenger transportation	66.12	65.91	75.19	82.24	82.40	77.29	76.07	63.96	58.27
Pipeline transportation	141.56	152.07	163.22	183.62	199.12	215.70	222.22	225.86	241.30
Other transportation and support activities	67.06	63.15	60.10	57.42	58.03	55.89	52.43	47.99	44.76
Warehousing and storage	36.88	36.01	35.37	35.50	35.16	35.26	33.41	32.19	31.02
Information	109.80	107.00	103.82	104.81	109.50	111.34	113.78	112.82	111.92
Publishing industries (includes software)	36.69	35.54	35.62	36.59	38.72	40.84	44.88	44.04	43.54
Motion picture and sound recording industries	151.56	150.37	148.42	146.06	151.70	149.62	148.78	156.91	161.29
Broadcasting and telecommunications	144.92	140.43	134.52	135.13	140.34	142.86	144.16	142.80	140.71
Information and data processing services	27.99	35.55	42.11	46.95	54.24	59.61	65.37	65.15	64.23
Finance and insurance	47.38	46.87	46.06	43.90	45.80	46.73	45.53	44.41	42.84
Federal Reserve banks, credit intermediation, and related activities	55.99	51.72	46.58	42.60	39.91	37.39	36.16	35.53	33.84
Securities, commodity contracts, and investments	27.77	25.35	23.28	21.20	25.43	28.64	27.01	24.93	23.02
Insurance carriers and related activities	31.59	31.50	31.96	31.36	31.47	31.15	30.02	28.03	25.71
Funds, trusts, and other financial vehicles	16.13	15.77	15.33	13.88	16.99	20.25	21.63	22.05	22.78
Real estate and rental and leasing	342.22	343.42	345.16	340.60	343.33	345.12	345.29	339.89	331.59
Real estate	371.75	375.67	377.46	372.90	373.19	371.66	373.15	366.14	356.17
Rental and leasing services and lessors of intangible assets	94.87	92.70	103.06	106.53	112.85	117.01	110.40	108.72	108.87
Professional, scientific, and technical services	22.23	22.42	23.88	24.81	25.98	27.52	28.14	27.87	27.87

	8.02	8.16	8.64	9.28	9.50	10.04	9.81	10.15	10.29
Legal services	22.59	22.08	23.87	25.30	27.50	28.67	28.36	28.92	27.13
Computer systems design and related services	26.43	26.85	28.45	29.02	30.33	32.47	34.01	33.18	33.49
Miscellaneous professional, scientific, and technical services	79.60	73.32	70.44	66.64	67.76	69.14	65.47	57.61	51.78
Management of companies and enterprises	26.00	24.97	25.44	25.74	27.73	29.36	29.67	29.05	27.79
Administrative and waste management services									
Administrative and support services	15.66	16.01	17.38	18.31	20.33	22.03	22.81	22.72	21.92
Waste management and remediation services	102.63	93.51	86.90	85.85	86.20	86.93	82.33	77.65	73.37
Educational services	85.80	89.35	91.50	93.89	94.79	99.04	101.48	105.63	106.32
Health care and social assistance	49.19	49.16	49.44	48.82	47.17	46.67	46.62	46.61	46.70
Ambulatory health care services	36.65	36.16	35.66	34.23	32.04	31.35	30.64	29.85	29.90
Hospitals	84.45	84.41	85.73	86.32	84.77	82.23	82.46	82.84	82.26
Nursing and residential care facilities	18.60	19.25	20.55	21.31	21.83	23.84	24.78	25.80	26.25
Social assistance	20.19	21.01	21.07	20.86	20.63	21.50	22.29	23.19	23.40
Arts, entertainment, and recreation	67.49	70.29	74.56	81.19	85.50	86.74	85.26	85.13	85.08
Performing arts, spectator sports, museums, and related activities	81.86	80.69	82.56	83.75	81.92	79.40	79.18	81.03	81.04
Amusements, gambling, and recreation industries	54.84	60.53	67.13	78.49	89.66	95.86	92.51	89.66	89.65
Accommodation and food services	45.32	45.67	45.97	44.66	45.94	45.61	45.38	43.77	43.18
Accommodation	79.05	80.34	81.10	76.87	85.16	88.50	89.36	86.08	84.33
Food services and drinking places	32.26	32.25	32.00	31.26	30.84	29.60	29.17	27.87	27.38
Other services, except government	44.95	44.71	46.17	47.50	49.29	50.74	53.26	53.95	55.36

Note: Asset Intensity is computed by dividing historical cost private fixed assets by the gross output in that service industry sector and is expressed as a percentage. Fixed assets and gross output data are sourced from U.S. Department of Commerce

Table 3.6b U.S. asset intensity (2006–2013)

Industry	2006	2007	2008	2009	2010	2011	2012	2013
Wholesale trade	30.55	29.18	29.08	34.75	30.73	29.73	30.14	30.49
Retail trade	50.93	52.59	56.09	58.94	54.67	53.68	50.81	49.63
Transportation and warehousing	69.30	68.01	68.15	80.94	75.82	71.99	72.37	72.90
Air transportation	104.62	100.13	98.06	117.68	103.58	93.87	95.20	
Railroad transportation	196.96	198.87	188.35	246.93	215.37	202.62	205.92	
Water transportation	65.88	59.01	53.56	59.96	54.81	52.76	54.31	
Truck transportation	35.20	34.04	33.34	39.45	36.39	35.11	35.63	
Transit and ground passenger transportation	52.25	51.56	50.33	50.11	50.33	46.68	44.55	
Pipeline transportation	249.59	272.20	286.73	379.58	376.72	371.43	385.76	
Other transportation and support activities	42.92	40.27	38.79	43.78	40.90	38.90	38.32	
Warehousing and storage	30.08	29.28	30.96	33.61	32.37	30.66	30.13	
Information	112.75	114.08	115.05	120.72	119.33	117.61	114.14	112.22
Publishing industries (includes software)	44.26	45.17	46.74	51.74	52.15	51.97	51.44	
Motion picture and sound recording industries	165.82	170.52	178.07	189.91	182.07	188.42	185.59	
Broadcasting and telecommunications	142.21	144.07	145.49	150.01	146.10	142.58	138.10	
Information and data processing services	62.55	62.85	57.57	58.79	65.58	66.79	62.39	
Finance and insurance	42.00	41.56	44.10	46.32	45.07	45.14	44.19	43.94
Federal Reserve banks, credit intermediation, and related activities	32.93	32.51	33.47	34.78	35.74	37.61	35.95	
Securities, commodity contracts, and investments	20.71	20.40	24.80	28.36	27.51	27.68	25.49	
Insurance carriers and related activities	24.59	23.37	23.52	25.47	23.68	22.19	21.51	
Funds, trusts, and other financial vehicles	22.32	21.70	22.06	26.92	25.00	23.43	24.49	
Real estate and rental and leasing	342.74	355.26	366.63	381.12	375.85	369.57	361.48	357.50
Real estate	367.76	381.87	398.00	409.12	405.60	399.75	391.29	
Rental and leasing services and lessors of intangible assets	116.72	118.96	110.08	119.88	111.14	108.84	110.57	
Professional, scientific, and technical services	27.67	27.05	27.32	29.37	28.96	27.87	27.61	27.77

Legal services	10.32	10.54	11.16	11.72	11.21	10.56	10.26	
Computer systems design and related services	26.83	24.29	23.06	24.91	23.11	21.16	19.42	
Miscellaneous professional, scientific, and technical services	33.08	32.47	32.80	35.40	35.39	34.33	34.51	
Management of companies and enterprises	47.20	43.13	44.89	49.51	46.50	44.22	41.86	40.45
Administrative and waste management services	27.57	26.87	26.55	29.72	28.64	27.78	27.28	27.99
Administrative and support services	21.90	21.41	21.36	24.21	23.60	22.70	22.30	
Waste management and remediation services	70.25	69.31	65.95	71.56	64.81	65.15	65.02	
Educational services	104.92	103.83	104.17	104.98	102.00	102.02	99.49	99.26
Health care and social assistance	47.47	48.66	49.24	49.69	50.05	50.05	49.69	49.54
Ambulatory health care services	29.73	30.09	29.82	29.41	29.42	28.85	28.27	
Hospitals	83.55	86.18	87.75	88.14	89.36	89.20	87.63	
Nursing and residential care facilities	26.91	27.76	28.78	29.64	30.02	30.71	31.22	
Social assistance	24.52	25.21	24.62	24.98	24.38	24.55	24.71	
Arts, entertainment, and recreation	81.62	80.16	81.64	86.01	86.50	84.41	81.27	80.38
Performing arts, spectator sports, museums, and related activities	77.48	76.62	76.86	78.68	79.88	79.27	76.53	
Amusements, gambling, and recreation industries	86.24	84.12	87.20	94.77	94.16	90.36	86.84	
Accommodation and food services	43.57	45.35	48.82	52.47	51.04	48.70	46.60	45.12
Accommodation	88.28	94.39	108.15	126.79	126.34	118.73	113.83	
Food services and drinking places	26.40	26.59	26.54	27.05	26.12	25.38	24.83	
Other services, except government	55.54	57.49	58.41	63.21	62.95	62.14	60.02	60.77

Note: Asset Intensity is computed by dividing historical cost of private fixed assets by the gross output in that service industry sector and is expressed as a percentage. Fixed assets and gross output data are sourced from U.S. Department of Commerce

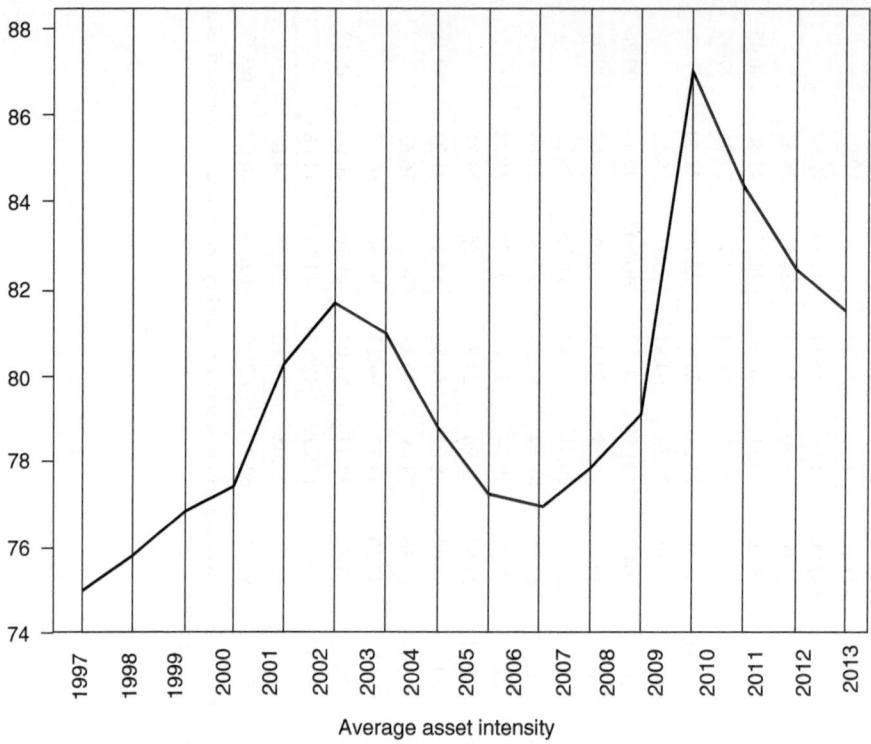

Fig. 3.8 U.S. asset intensity for the years 1997–2012

3.8   Labor Ratio

Labor Ratio measures the cost of labor faced by a firm on its outputs. A higher value indicates the production of a particular service is very labor-intensive or requires high cost labor to produce services. This variable was calculated as follows: First, total compensation costs of the employees and gross output in a particular service sector were obtained from various U.S. government websites. This ratio was then computed by dividing compensation cost of employees by the gross output in that service industry sector and is expressed as a percentage. Table 3.7a presents Labor Ratio for the United States for the years 1998–2005 and Table 3.7b presents the same for the years 2006–2013. Figure 3.9 illustrates the average labor ratio across service industries for 1998–2012, which show a declining trend in recent years, the exception being 2009. In the years 1998–2013, the service industry average Labor Ratio had values ranging from 36.58 to 33.18, with an average of 34.61.

Table 3.7a U.S. labor ratio (1998–2005)

Industry	1998	1999	2000	2001	2002	2003	2004	2005
Wholesale trade	37.37	36.51	37.08	38.28	36.81	36.40	34.91	33.96
Retail trade	40.92	40.15	40.87	42.11	41.18	40.26	38.73	38.71
Motor vehicle and parts dealers	40.07	37.85	40.02	46.46	40.68	41.70	41.51	44.05
Food and beverage stores	46.60	45.37	46.92	44.88	45.57	45.44	43.35	42.23
General merchandise stores	40.43	39.33	41.48	42.01	44.41	41.97	39.82	40.23
Other retail	39.72	39.79	39.40	39.89	39.29	37.86	36.25	35.69
Transportation and warehousing	34.45	34.62	34.57	35.70	35.10	33.76	32.33	30.88
Air transportation	30.01	32.27	32.02	38.98	40.49	35.36	31.85	27.33
Rail transportation	42.20	43.11	41.54	40.31	38.88	38.68	36.44	34.02
Water transportation	12.26	12.34	11.83	12.94	13.68	12.18	11.23	11.88
Truck transportation	30.16	30.85	30.00	30.29	29.70	29.46	28.08	27.00
Transit and ground passenger transportation	34.88	39.06	40.49	40.30	40.63	40.92	36.74	34.49
Pipeline transportation	14.72	15.88	21.19	21.51	16.96	16.54	16.03	16.37
Other transportation and support activities	46.37	43.34	42.47	41.35	39.89	39.41	39.03	38.24
Warehousing and storage	64.21	55.89	59.12	55.39	51.79	48.27	51.16	52.31
Information	23.57	24.23	24.58	24.05	21.94	22.05	22.03	21.49
Publishing industries (includes software)	28.57	29.66	29.95	29.33	26.57	27.66	26.39	26.22
Motion picture and sound recording industries	20.51	19.21	18.50	20.01	18.17	17.49	18.96	18.59
Broadcasting and telecommunications	19.18	19.54	19.30	18.67	17.95	17.98	18.09	17.38
Information and data processing services	46.39	50.09	55.44	52.77	40.48	39.39	38.45	37.23
Finance and insurance	30.25	29.20	28.40	30.62	30.03	29.48	29.78	29.61
Federal Reserve banks, credit intermediation, and related activities	26.60	25.33	23.88	25.09	25.10	26.53	27.22	27.70
Securities, commodity contracts, and investments	42.64	40.14	40.00	48.42	47.25	41.99	43.39	43.23
Insurance carriers and related activities	31.68	31.44	30.42	30.94	30.62	29.60	28.58	27.43
Funds, trusts, and other financial vehicles	1.45	1.34	1.19	1.43	1.60	1.76	1.82	1.83
Real estate and rental and leasing	4.58	4.62	4.60	4.48	4.48	4.26	4.22	4.06
Real estate	3.57	3.63	3.62	3.53	3.55	3.41	3.40	3.29
Rental and leasing services and lessors of intangible assets	12.39	12.05	11.69	11.85	12.54	11.42	11.49	11.07

(continued)

Table 3.7a   (continued)

Industry	1998	1999	2000	2001	2002	2003	2004	2005
Professional, scientific, and technical services	40.84	41.02	42.75	42.67	41.54	40.66	40.49	41.10
Legal services	37.69	37.18	39.15	39.49	39.42	37.53	37.39	36.75
Computer systems design and related services	51.48	51.06	55.17	54.72	51.00	50.76	53.34	54.85
Miscellaneous professional, scientific, and technical services	38.96	39.19	40.00	40.11	39.61	39.06	38.33	39.14
Management of companies and enterprises	53.41	52.57	56.65	57.87	57.27	56.74	54.91	52.40
Administrative and waste management services	44.52	44.08	46.01	47.52	48.79	47.95	47.65	47.41
Administrative and support services	46.60	46.22	48.26	49.86	51.18	50.36	50.13	49.83
Waste management and remediation services	28.66	27.70	27.78	28.96	30.09	29.42	28.72	28.59
Educational services	51.94	52.24	51.70	50.59	53.29	53.48	55.69	54.67
Health care and social assistance	50.30	50.67	50.41	49.73	49.54	49.65	50.02	49.58
Ambulatory health care services	48.19	48.30	48.24	47.53	47.38	47.01	47.32	47.41
Hospitals	51.11	51.41	50.95	51.07	50.09	50.85	51.31	50.29
Nursing and residential care facilities	55.72	57.19	57.65	57.25	58.45	57.82	57.71	55.55
Social assistance	51.57	52.10	50.53	46.83	47.63	48.78	49.90	50.81
Arts, entertainment, and recreation	34.56	35.18	36.98	37.18	37.19	35.94	35.39	34.31
Performing arts, spectator sports, museums, and related activities	33.56	34.50	35.20	33.79	32.77	32.10	32.34	31.17
Amusements, gambling, and recreation industries	35.50	35.81	38.73	40.98	42.58	40.52	38.85	37.86
Accommodation and food services	33.55	34.76	34.11	34.95	34.43	34.73	34.89	34.47
Accommodation	33.58	34.51	33.10	34.78	33.58	33.14	32.98	31.90
Food services and drinking places	33.54	34.85	34.53	35.01	34.75	35.31	35.61	35.46
Other services, except government	39.98	39.77	40.56	40.19	41.37	41.94	42.35	42.22

Note: Labor ratio is computed by dividing compensation cost of employees by the gross output in that service industry sector and is expressed as a percentage. Labor ratio and gross output data are sourced from U.S. Department of Labor and U.S. Department of Commerce respectively

Table 3.7b U.S. labor ratio (2006–2013)

Industry	2006	2007	2008	2009	2010	2011	2012	2013
Wholesale trade	33.83	33.81	33.21	37.11	32.92	32.24	32.38	31.65
Retail trade	38.01	38.32	39.08	38.82	36.12	36.12	34.62	34.21
Motor vehicle and parts dealers	44.29	44.91	48.09	50.95	39.51	41.19	37.15	
Food and beverage stores	40.74	41.41	41.46	41.34	40.34	38.91	38.11	
General merchandise stores	40.73	40.21	39.91	39.40	37.34	39.33	37.98	
Other retail	34.71	35.08	35.76	35.13	33.56	32.95	31.90	
Transportation and warehousing	29.47	29.51	28.14	31.19	29.09	28.08	28.47	28.14
Air transportation	25.40	25.58	23.25	25.86	23.97	22.90	23.52	
Rail transportation	31.54	31.16	29.34	33.99	27.52	26.59	27.71	
Water transportation	12.06	11.90	11.12	11.97	11.11	11.29	11.51	
Truck transportation	26.77	27.12	26.05	29.16	27.28	26.57	26.44	
Transit and ground passenger transportation	32.07	31.86	32.72	33.26	33.89	34.07	33.18	
Pipeline transportation	17.34	18.27	16.24	20.45	20.23	20.46	23.35	
Other transportation and support activities	36.00	35.90	34.98	38.33	36.71	34.89	34.77	
Warehousing and storage	46.45	45.29	43.82	45.49	42.84	40.98	42.63	
Information	21.35	21.55	20.78	20.92	19.84	19.93	19.70	19.90
Publishing industries (includes software)	26.70	28.26	27.96	28.82	28.35	29.36	30.48	
Motion picture and sound recording industries	18.61	19.42	20.49	21.31	20.54	21.60	21.69	
Broadcasting and telecommunications	17.04	18.05	17.09	17.05	15.68	15.29	14.67	
Information and data processing services	34.99	25.86	22.98	22.63	22.01	21.86	21.14	
Finance and insurance	29.97	29.99	30.96	29.61	30.29	31.81	31.57	30.98
Federal Reserve banks, credit intermediation, and related activities	28.49	28.44	27.88	26.80	27.93	31.16	31.64	
Securities, commodity contracts, and investments	42.15	43.57	50.81	44.05	46.61	49.01	46.70	
Insurance carriers and related activities	27.59	26.77	27.11	28.08	27.34	27.60	27.74	
Funds, trusts, and other financial vehicles	1.76	0.55	0.53	0.61	0.49	0.54	0.45	
Real estate and rental and leasing	4.14	4.22	4.16	4.02	3.93	3.93	4.05	4.06
Real estate	3.37	3.42	3.39	3.25	3.22	3.22	3.32	
Rental and leasing services and lessors of intangible assets	11.16	11.39	10.47	11.19	10.29	10.12	10.14	

(continued)

Table 3.7b (continued)

Industry	2006	2007	2008	2009	2010	2011	2012	2013
Professional, scientific, and technical services	42.42	42.37	42.83	43.19	42.40	42.98	44.38	44.74
Legal services	37.27	38.12	39.35	39.73	38.80	39.09	39.27	
Computer systems design and related services	57.43	56.33	55.09	55.40	53.98	55.59	58.98	
Miscellaneous professional, scientific, and technical services	40.38	40.13	40.62	40.91	40.15	40.44	41.49	
Management of companies and enterprises	52.12	50.83	50.20	49.98	50.77	49.19	49.06	48.38
Administrative and waste management services	47.02	46.67	46.02	45.76	45.69	45.15	45.94	47.60
Administrative and support services	49.60	49.16	48.49	47.96	47.37	47.57	48.12	
Waste management and remediation services	27.52	27.29	27.28	29.02	33.73	27.41	29.42	
Educational services	53.90	53.15	52.84	53.88	51.96	52.23	51.27	50.05
Health care and social assistance	50.08	50.09	50.23	50.28	49.98	49.73	49.42	49.27
Ambulatory health care services	48.56	48.71	48.84	49.10	49.57	49.79	50.21	
Hospitals	50.40	50.47	50.60	50.24	48.82	48.48	46.98	
Nursing and residential care facilities	55.63	55.29	55.89	56.28	56.27	55.03	54.50	
Social assistance	49.83	49.19	48.94	49.14	48.86	47.91	49.15	
Arts, entertainment, and recreation	33.25	32.22	32.64	32.35	32.43	32.09	32.10	32.18
Performing arts, spectator sports, museums, and related activities	31.57	30.72	30.86	30.44	30.77	30.76	30.78	
Amusements, gambling, and recreation industries	35.14	33.89	34.72	34.64	34.38	33.61	33.63	
Accommodation and food services	34.19	34.94	34.90	34.69	34.57	34.63	34.79	34.14
Accommodation	31.31	32.21	33.12	33.51	34.01	33.93	34.05	
Food services and drinking places	35.30	35.97	35.58	35.10	34.76	34.85	35.04	
Other services, except government	41.93	43.30	43.51	44.84	43.12	43.44	43.51	45.06

Note: Labor ratio is computed by dividing compensation cost of employees by the gross output in that service industry sector and is expressed as a percentage. Labor ratio and gross output data are sourced from U.S. Department of Labor and U.S. Department of Commerce respectively

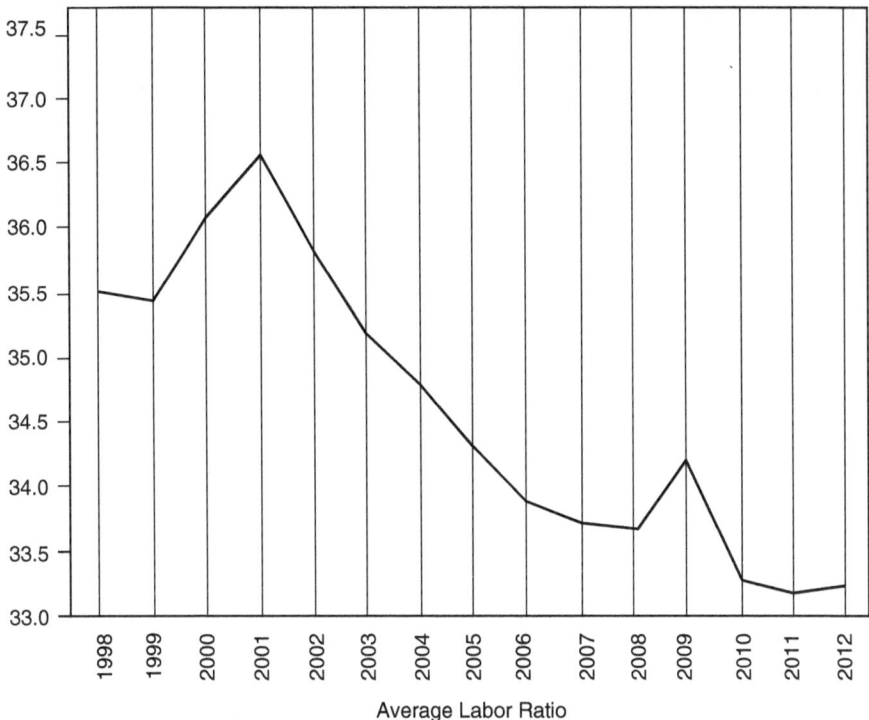

Fig. 3.9 U.S. labor ratio for the years 1998–2012

3.9   Profit Margin Before Tax

Profit Margin Before Tax is profits made per unit of sales. This variable offers one of the important indicators of profitability, as it represents cash left after payment of all expenses, before taxes. This variable was computed by the author as follows: Initially, corporate profits before tax and gross output figures were collected for each service industry segment from U.S. government sources. Profit Margin was then computed by dividing corporate profits before tax by the gross output in that service industry sector and is expressed as a percentage. Table 3.8a presents Profit Margin Before Tax for the years 1998–2004 and Table 3.8b presents the same for the years 2004–2012. Figure 3.10 shows the average Profit Margin Before Tax across service industries for the years 1998–2011, while Figure 3.11 shows the total profits before tax for the service sector for same years (It should be noted the table presents data up to 2012. However, since there are many missing entries the figure focuses only to 2011.) During the period 1998–2011, the average profit before taxes in the service sector was 4.07 %, varying between 1.34 % in 2000 and 6.33 % in 2006. In aggregate terms, total profit before taxes for service firms, which was at $526,947 million in 1998, rose to $1,412,507 million in 2011.

Table 3.8a   U.S. profit margin before tax (1998–2004)

Industry	1998	1999	2000	2001	2002	2003	2004
Wholesale trade	6.58	6.80	7.11	5.59	6.05	6.81	8.21
Retail trade	6.96	6.51	5.39	7.03	7.67	8.12	8.35
Transportation and warehousing	2.40	1.38	1.65	−0.17	−0.97	0.79	1.80
Air transportation	3.94	2.38	−0.35	−11.14	−16.49	−7.37	−6.98
Rail transportation	1.46	0.49	4.96	2.82	3.11	3.30	5.23
Water transportation	0.99	−0.75	1.46	2.13	0.92	1.05	1.38
Truck transportation	1.81	1.24	1.51	1.37	0.85	1.98	2.94
Transit and ground passenger transportation	2.59	1.58	2.37	2.58	3.32	2.50	2.44
Pipeline transportation	4.17	2.27	−0.14	3.70	4.11	3.20	4.33
Other transportation and support activities	1.96	0.83	3.07	3.70	4.17	4.03	5.92
Warehousing and storage	2.41	2.51	1.89	1.50	1.72	1.02	1.40
Information	4.09	2.25	−1.20	−2.59	−0.38	1.50	4.84
Publishing industries (includes software)	5.15	6.81	0.84	0.92	4.35	5.57	7.39
Motion picture and sound recording industries	1.12	1.67	0.76	3.13	6.68	5.82	8.62
Broadcasting and telecommunications	5.25	1.35	0.04	−3.53	−4.16	−1.30	2.22
Information and data processing services	−4.98	−7.00	−20.72	−15.71	0.86	1.61	8.77
Finance and insurance	7.33	6.74	4.34	6.61	10.66	12.11	12.60
Federal Reserve banks	5.52	5.18	5.55	4.80	3.67	3.03	2.97
Securities, commodity contracts, and investments	−0.59	−0.56	−6.63	−1.87	2.44	3.98	5.22
Insurance carriers and related activities	6.90	4.89	3.53	1.85	5.33	10.61	15.27
Funds, trusts, and other financial vehicles	−23.98	−20.66	−23.71	−4.29	16.31	20.61	8.90
Real estate and rental and leasing	0.68	0.61	0.52	0.39	0.47	0.47	0.54
Real estate	0.47	0.52	0.54	0.32	0.44	0.48	0.65
Rental and leasing services and lessors of intangible assets	2.32	1.24	0.42	0.90	0.71	0.34	−0.37
Professional, scientific, and technical services	3.12	2.81	0.82	1.10	2.13	2.72	3.20
Legal services	1.56	2.23	3.10	4.14	4.25	3.77	3.86
Computer systems design and related services	4.65	1.97	−6.32	−4.86	−1.20	1.11	2.17

Table 3.8a (continued)

Industry	1998	1999	2000	2001	2002	2003	2004
Miscellaneous professional, scientific, and technical services	3.19	3.23	2.34	1.95	2.40	2.80	3.24
Management of companies and enterprises	24.32	28.27	33.29	38.07	43.10	41.74	43.34
Administrative and waste management services	2.05	2.06	1.80	2.15	2.24	2.86	2.96
Administrative and support services	1.81	1.95	1.72	2.05	2.01	2.50	2.73
Waste management and remediation services	3.91	2.88	2.44	2.95	4.09	5.63	4.73
Educational services	1.55	1.50	1.40	1.09	1.40	1.76	2.38
Health care and social assistance	1.51	2.08	2.34	2.75	3.36	3.43	3.39
Ambulatory health care services	2.49	3.29	3.59	4.53	5.10	5.20	5.09
Hospitals and nursing and residential care facilities	0.50	0.82	0.76	0.56	1.30	1.34	1.35
Social assistance	1.41	2.09	3.71	4.30	4.57	4.64	4.65
Arts, entertainment, and recreation	2.19	2.10	1.68	2.08	2.48	2.44	3.16
Performing arts, spectator sports, museums, and related activities	1.87	1.99	1.50	1.92	2.10	2.55	3.31
Amusements, gambling, and recreation industries	2.49	2.21	1.86	2.26	2.94	2.31	2.99
Accommodation and food services	2.10	2.52	2.81	2.23	2.49	2.19	2.68
Accommodation	1.66	2.41	2.67	1.02	1.73	0.97	2.31
Food services and drinking places	2.27	2.56	2.87	2.70	2.78	2.64	2.82
Other services, except government	1.01	0.89	1.37	1.57	1.08	1.24	1.35

Note: Profit Margin is computed by dividing corporate profits before tax by the gross output in that service industry sector and is expressed as a percentage. Corporate profits before tax and gross output are sourced from U.S. Department of Commerce respectively

Table 3.8b U.S. profit margin before tax (2005–2012)

Industry	2005	2006	2007	2008	2009	2010	2011	2012
Wholesale trade	9.26	9.73	9.28	7.54	8.16	9.36	8.58	10.00
Retail trade	10.29	10.51	9.55	6.82	9.04	9.44	9.18	10.95
Transportation and warehousing	3.75	5.03	2.83	3.04	2.97	5.34	3.34	5.36
Air transportation	−4.67	1.83	−0.67	−1.64	0.38	7.06	3.59	
Rail transportation	12.95	17.28	16.22	13.50	10.16	13.02	9.76	
Water transportation	3.14	4.25	2.49	3.19	0.35	1.34	−0.24	
Truck transportation	4.16	3.64	2.17	2.34	2.13	3.66	2.21	
Transit and ground passenger transportation	2.32	2.21	1.50	0.96	2.14	3.36	2.60	
Pipeline transportation	4.54	3.74	5.11	5.01	7.21	4.55	−16.81	
Other transportation and support activities	8.17	7.82	2.79	5.14	4.73	6.68	7.28	
Warehousing and storage	2.39	2.46	1.76	1.55	2.04	2.09	1.65	
Information	8.13	9.21	9.00	7.49	6.76	7.60	6.42	7.35
Publishing industries (includes software)	9.69	7.02	6.17	4.57	4.77	7.05	6.85	
Motion picture and sound recording industries	4.67	6.31	7.57	9.58	10.02	14.68	16.36	
Broadcasting and telecommunications	8.38	10.65	11.08	8.49	7.25	6.95	5.04	
Information and data processing services	6.72	10.39	6.55	7.14	5.50	5.03	3.35	
Finance and insurance	13.72	13.04	8.10	−2.21	11.58	11.60	10.74	15.77
Federal Reserve banks	3.75	4.64	4.88	4.89	6.86	10.66	11.91	
Securities, commodity contracts, and investments	6.33	6.37	−3.02	−22.63	7.96	4.99	1.03	
Insurance carriers and related activities	15.89	16.17	12.10	−1.72	7.57	8.37	4.62	
Funds, trusts, and other financial vehicles	11.25	11.16	3.54	−7.42	−1.76	−4.60	−11.19	
Real estate and rental and leasing	1.08	0.90	0.69	0.22	0.28	0.51	0.09	0.79
Real estate	0.83	0.55	0.36	0.10	0.10	0.32	0.26	
Rental and leasing services and lessors of intangible assets	3.35	4.02	3.62	1.27	1.97	2.12	−1.39	
Professional, scientific, and technical services	3.52	3.95	3.77	4.00	4.43	5.16	4.58	4.40
Legal services	4.59	4.20	4.00	4.79	5.48	5.51	5.69	
Computer systems design and related services	1.61	2.85	3.14	3.56	5.21	5.09	4.32	

Table 3.8b   (continued)

Industry	2005	2006	2007	2008	2009	2010	2011	2012
Miscellaneous professional, scientific, and technical services	3.66	4.15	3.86	3.90	3.93	5.08	4.36	
Management of companies and enterprises	45.79	42.37	31.18	31.67	35.67	41.67	35.35	32.66
Administrative and waste management services	4.08	3.32	3.93	3.43	3.62	4.23	3.90	4.18
Administrative and support services	3.74	3.56	3.38	3.08	2.93	3.45	3.46	
Waste management and remediation services	6.70	1.51	8.25	6.02	8.87	9.77	7.16	
Educational services	2.70	2.53	2.30	2.41	3.70	4.08	3.23	2.67
Health care and social assistance	3.91	3.97	3.72	3.73	4.27	4.39	4.23	4.50
Ambulatory health care services	5.74	5.93	5.48	5.65	6.52	6.48	6.16	
Hospitals and nursing and residential care facilities	1.66	1.61	1.23	1.30	1.49	1.64	1.81	
Social assistance	5.82	5.93	7.43	6.38	7.24	8.01	7.13	
Arts, entertainment, and recreation	4.02	3.74	2.98	2.58	2.75	3.11	3.89	4.20
Performing arts, spectator sports, museums, and related activities	3.76	4.02	3.45	3.11	3.07	3.01	4.42	
Amusements, gambling, and recreation industries	4.32	3.44	2.44	1.96	2.38	3.23	3.27	
Accommodation and food services	3.44	3.55	3.01	1.99	1.91	2.47	3.02	3.47
Accommodation	3.27	3.40	2.38	1.01	−2.36	−1.60	1.52	
Food services and drinking places	3.51	3.61	3.25	2.36	3.37	3.82	3.51	
Other services, except government	2.26	2.32	2.07	1.29	1.83	2.79	2.78	2.85

Note: Profit Margin is computed by dividing corporate profits before tax by the gross output in that service industry sector and is expressed as a percentage. Corporate profits before tax and gross output are sourced from U.S. Department of Commerce respectively

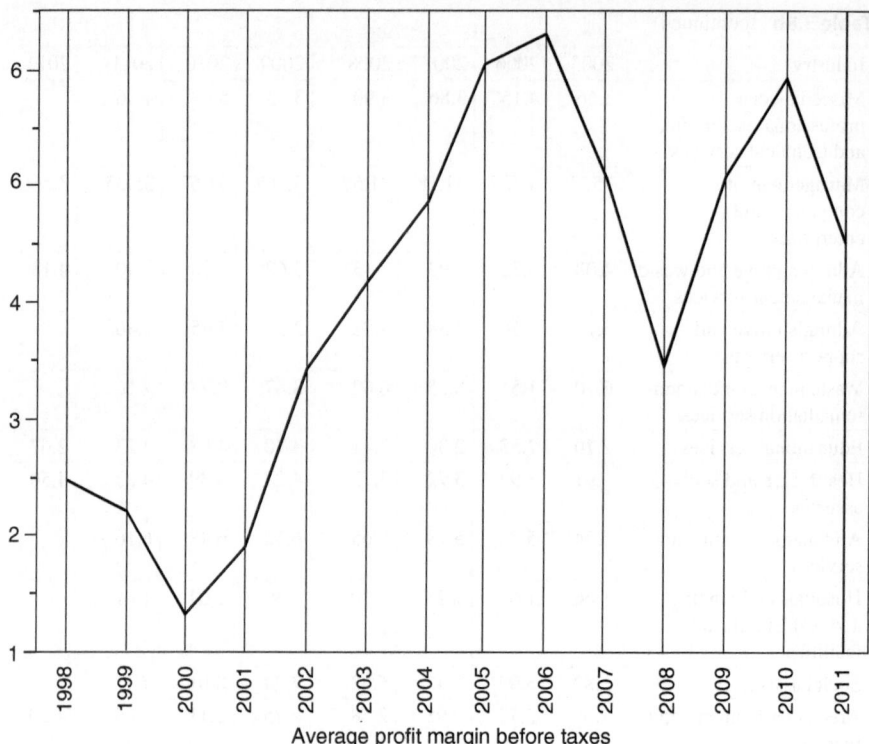

Fig. 3.10   U.S. average profit margin before tax for the years 1998–2011

Fig. 3.11 U.S. profits before tax for the years 1998–2011

3.10 Industry Size

Service industry sector size for the United States is presented in Table 3.9a for the years 2000–2006 and in Table 3.9b for the years 2006–2013, in billions of dollars. All data reported in the tables were collected from the Bureau of Economic Analysis, U.S. Department Commerce. Figure 3.12 presents the total service sector size for the Years 2000–2012 and Figure 3.13 presents the U.S. service sector growth rate for the Years 2001–2012. Current service sector output is at $42344.6 billion in 2012, which has risen from $26358.2 billion in 2000. As indicated in both figures, the service sector as a whole has been growing consistently, with the exception of the year 2009.

Table 3.9a   U.S. industry size (2000–2006) [Billions of U.S. dollars]

Industry	2000	2001	2002	2003	2004	2005	2006
Wholesale trade	884.3	862.6	893.9	925.8	1023	1116.8	1196.5
Retail trade	992.8	989.4	1040.4	1104.6	1185.5	1233.7	1295.1
Motor vehicle and parts dealers	194.3	175.7	208.5	211	216.8	211.4	214
Food and beverage stores	140.7	151.9	153.4	158.3	167.6	174.4	184.2
General merchandise stores	134.6	137.6	138.2	155.9	171.2	179.5	184.3
Other retail	523.2	524.2	540.3	579.3	629.9	668.4	712.6
Transportation and warehousing	590.9	582.5	587.5	620.7	686.4	749.9	818
Air transportation	123.7	107.2	100.7	109.9	120.5	133.6	145.1
Rail transportation	40.1	40.7	41.3	42.3	46.6	52.6	59.3
Water transportation	27.6	27.7	27	31.3	37.3	39.3	42.2
Truck transportation	201.3	202.1	206	211.9	234.3	259.7	276.7
Transit and ground passenger transportation	25.9	26.7	27.3	28	33.3	36.9	42.3
Pipeline transportation	23.2	22.7	22.3	22.5	23.2	23	24.4
Other transportation and support activities	115.3	118.9	123.1	129.7	141.5	151.7	166.6
Warehousing and storage	33.8	36.4	39.7	45.2	49.7	53.2	61.5
Information	1000.9	1027.9	1041.2	1045.1	1080.1	1123.2	1165
Publishing industries, except internet (includes software)	254.2	259.3	262	258.7	273.4	283.2	292.6
Motion picture and sound recording industries	110.3	112	119.5	127.3	127.4	131.5	134.3
Broadcasting and telecommunications	562.7	579.8	577	574.5	587.1	610.1	627.3
Data processing, internet publishing, and other information services	73.7	76.7	82.7	84.6	92.1	98.4	110.8
Finance, insurance, real estate, rental, and leasing	3167.2	3266.8	3396.6	3612.5	3922.9	4305.1	4538.2
Finance and insurance	1450.5	1455.1	1478.3	1562.2	1674	1807.9	1934.4
Federal Reserve banks, credit intermediation, and related activities	563.1	601.1	641.6	663.4	675.2	708.9	728.5
Securities, commodity contracts, and investments	357.1	313.8	287.4	309.2	345.7	383.2	448.5
Insurance carriers and related activities	438.8	457.9	476.8	518.4	571.1	625.9	657.5
Funds, trusts, and other financial vehicles	91.5	82.4	72.6	71.2	82.1	90	99.9
Real estate and rental and leasing	1716.8	1811.6	1918.3	2050.3	2248.9	2497.1	2603.8

Table 3.9a   (continued)

Industry	2000	2001	2002	2003	2004	2005	2006
Real estate	1508.6	1603.8	1718.4	1832.9	2019.6	2249	2344.3
Rental and leasing services and lessors of intangible assets	208.2	207.8	199.9	217.4	229.3	248.1	259.5
Professional and business services	1818.1	1849.2	1856.6	1945.3	2099	2276.8	2436.2
Professional, scientific, and technical services	1082.2	1118.2	1127.3	1173.1	1248.8	1344.4	1429.9
Legal services	191.9	205.3	216.1	239.5	251.3	268.3	279.1
Computer systems design and related services	206.7	204.7	194.3	192.2	195.4	208.6	221.8
Miscellaneous professional, scientific, and technical services	683.6	708.3	717	741.3	802.1	867.4	929
Management of companies and enterprises	260.5	259.6	262.5	281.2	319.4	353.2	384.5
Administrative and waste management services	475.5	471.4	466.9	491	530.9	579.3	621.8
Administrative and support services	423.2	418.5	414	434.4	469.6	513.2	549.2
Waste management and remediation services	52.3	52.9	52.8	56.6	61.3	66.1	72.6
Educational services, health care, and social assistance	1113.6	1212.7	1308.7	1397.3	1482.8	1579.8	1673.6
Educational services	144	157.4	166.4	176.1	182.9	193	209.3
Health care and social assistance	969.6	1055.4	1142.2	1221.3	1299.9	1386.8	1464.3
Ambulatory health care services	453.7	494.1	531.1	568.6	609.8	645.2	674.3
Hospitals	320.8	346.7	382.6	409.4	435.9	470.6	503.2
Nursing and residential care facilities	111.2	120	127.1	136	143.4	153.5	162
Social assistance	83.9	94.5	101.4	107.2	110.8	117.5	124.8
Arts, entertainment, recreation, accommodation, and food services	662.8	679.9	712.9	750.7	800.1	846.1	902.4
Arts, entertainment, and recreation	149.4	155.9	165.2	178.4	189.7	199.7	219.8
Performing arts, spectator sports, museums, and related activities	75.1	82.4	90.3	97	100.7	106	115.9
Amusements, gambling, and recreation industries	74.4	73.5	74.8	81.4	89	93.7	103.9
Accommodation and food services	513.4	524	547.7	572.3	610.5	646.4	682.6

(continued)

Table 3.9a (continued)

Industry	2000	2001	2002	2003	2004	2005	2006
Accommodation	150.9	145.6	148.7	154.1	166.7	179.3	189.4
Food services and drinking places	362.5	378.4	399	418.2	443.8	467.1	493.1
Other services, except government	423.6	441.1	457	468.1	486.6	496.2	522.3

Data Source: U.S. Department of Commerce

Table 3.9b U.S. industry size (2007–2013) [Billions of U.S. dollars]

Industry	2007	2008	2009	2010	2011	2012	2013
Wholesale trade	1269.5	1311.4	1092	1253.1	1354.4	1413.1	1477.4
Retail trade	1320.7	1280.7	1222.5	1327.3	1374.8	1478.1	1547
Motor vehicle and parts dealers	214.2	187.2	153.2	206.6	212	249.1	
Food and beverage stores	186.9	194.3	191.9	198.9	208.1	217.9	
General merchandise stores	193.9	200.1	208.7	216.6	215.2	223.2	
Other retail	725.6	699.2	668.7	705.3	739.5	787.8	
Transportation and warehousing	866.9	907.9	773.5	844.8	925.9	965.3	1004.4
Air transportation	156	165	136.3	153.6	171.4	172.8	
Rail transportation	61.9	69.5	55.4	67	76.2	79.4	
Water transportation	48.3	54.7	49.2	54	56.1	56.9	
Truck transportation	283.2	287.6	233.2	253.4	281.4	299.5	
Transit and ground passenger transportation	44.8	45.3	45.1	45.5	48.2	51.4	
Pipeline transportation	25.9	29.4	24	26.2	28	28.8	
Other transportation and support activities	177.8	183.3	159.2	168.7	183.8	193.1	
Warehousing and storage	69	73	71.1	76.3	80.9	83.3	
Information	1208.6	1240.4	1202.7	1252.8	1305.8	1377.8	1437.7
Publishing industries, except internet (includes software)	304.4	308.1	286.6	288.2	297.3	303.1	
Motion picture and sound recording industries	135.7	132.7	125.9	136.1	134.7	140.2	
Broadcasting and telecommunications	648.7	666.9	657.7	687.4	717.5	757.4	
Data processing, internet publishing, and other information services	119.8	132.7	132.5	141.2	156.3	177.1	
Finance, insurance, real estate, rental, and leasing	4708.4	4620.1	4425.3	4533.7	4624.8	4831.8	5016.5
Finance and insurance	2061.6	1979.9	1854.3	1896	1907.7	1999.4	2073.1

Table 3.9b (continued)

Industry	2007	2008	2009	2010	2011	2012	2013
Federal Reserve banks, credit intermediation, and related activities	738	716.7	689.8	671.3	637.9	667.4	
Securities, commodity contracts, and investments	503.4	432.7	403	418.4	413.3	436.6	
Insurance carriers and related activities	703.6	710.4	662.4	699.4	742.1	776.9	
Funds, trusts, and other financial vehicles	116.6	120.1	99.2	106.8	114.4	118.4	
Real estate and rental and leasing	2646.8	2640.2	2571	2637.7	2717.1	2832.4	2943.4
Real estate	2378.9	2352.5	2322.1	2371.1	2435.3	2531.6	
Rental and leasing services and lessors of intangible assets	267.9	287.8	249	266.6	281.8	300.8	
Professional and business services	2639	2737.8	2596	2727.3	2889.3	3027.8	3105.8
Professional, scientific, and technical services	1543.1	1616.4	1548.7	1609.2	1691.9	1750.6	1789.3
Legal services	288.5	287.7	276.4	280.1	287.9	294.4	
Computer systems design and related services	249.5	271.5	266.2	289.9	310	327	
Miscellaneous professional, scientific, and technical services	1005.1	1057.2	1006.1	1039.1	1093.9	1129.2	
Management of companies and enterprises	431.7	439.3	415.3	447.1	483.7	531.1	568.8
Administrative and waste management services	664.2	682	632	671	713.8	746.1	747.7
Administrative and support services	588.6	602.5	558.5	588.6	628.3	659.2	
Waste management and remediation services	75.6	79.6	73.5	82.4	85.5	86.9	
Educational services, health care, and social assistance	1774.7	1886	1975.3	2058.6	2147.6	2249.3	2339.4
Educational services	227.1	244.4	259	279.4	291.7	311.7	324.3
Health care and social assistance	1547.6	1641.6	1716.2	1779.1	1855.9	1937.6	2015.1
Ambulatory health care services	710.9	750.9	777.3	805.2	836.8	869.6	
Hospitals	531.1	566.5	603	623.4	657.2	698.5	
Nursing and residential care facilities	171.5	180.7	186.6	194.2	202.2	207.9	
Social assistance	134.1	143.4	149.3	156.3	159.7	161.5	

(continued)

Table 3.9b   (continued)

Industry	2007	2008	2009	2010	2011	2012	2013
Arts, entertainment, recreation, accommodation, and food services	948	969.6	943.1	964	1013.4	1075.3	1130
Arts, entertainment, and recreation	238.9	246.7	243.1	245.2	254.7	268.6	277.3
Performing arts, spectator sports, museums, and related activities	126.2	131.8	132.3	132.2	136.5	144	
Amusements, gambling, and recreation industries	112.7	114.8	110.8	113	118.2	124.6	
Accommodation and food services	709	723	700	718.8	758.7	806.8	852.6
Accommodation	196.1	197.6	178.4	178.8	189.5	197.4	
Food services and drinking places	513	525.3	521.6	539.9	569.3	609.3	
Other services, except government	533.8	551.6	526	539.6	553.9	579.6	579.4

Data Source: U.S. Department of Commerce

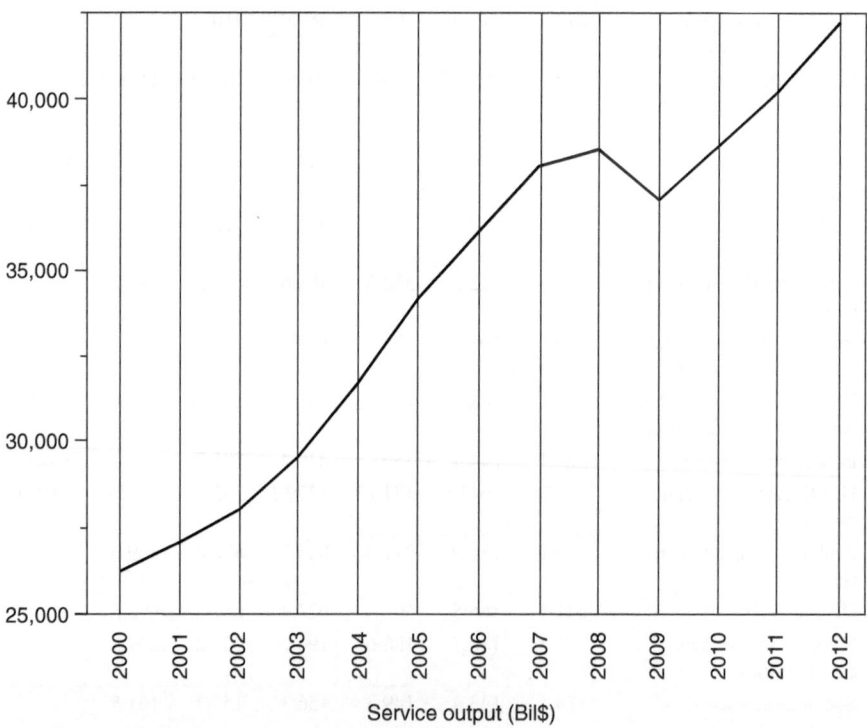

Fig. 3.12   U.S. industry size for the years 2000–2012

Fig. 3.13 U.S. service sector growth rate for the years 2001–2012

Data Sources

United States Department of Commerce. http://www.bea.gov/
United States Department of Labor. http://www.bls.gov/

Service Sector Data on 15 OECD Nations

4

4.1 Export Ratio

Export Ratio represents the ratio of value of services that was transferred outside a country's borders relative to the level of economic output that takes place in a particular industry in a country. Stated differently, it represents the extent of exports in an industry. This ratio is illustrative as to the extent to which a country's industries are successful in exporting products, thereby serving as a proxy for the relative strength of a specific country in the particular service sector. Export Ratio numbers were derived from the OECD database as follows: Initially, exports in a particular service sector and the gross productive output were determined from OECD databases for each service industry segment in a country. Then, exports from the country were divided by the gross productive output in the corresponding service sector and multiplied by 100 to arrive at the Export Ratio.

Export Ratio numbers for the years 2000–2010 for 15 OECD countries across 11–12 industries are presented where data was available. Table 4.1 shows the Export Ratio for Austria; Table 4.2 shows the Export Ratio for Belgium; Table 4.3 shows the Export Ratio for the Czech Republic; Table 4.4 shows the Export Ratio for Denmark; Table 4.5 shows the Export Ratio for Finland; Table 4.6 shows the Export Ratio for France; Table 4.7 shows the Export Ratio for Germany; Table 4.8 shows the Export Ratio for Hungary; Table 4.9 shows the Export Ratio for Italy; Table 4.10 shows the Export Ratio for Korea; Table 4.11 shows the Export Ratio for the Netherlands; Table 4.12 shows the Export Ratio for Norway; Table 4.13 shows the Export Ratio for Slovenia; Table 4.14 shows the Export Ratio for Sweden; and Table 4.15 shows the Export Ratio for the United States. Figure 4.1 graphically shows the average Export Ratio for the services sector in each of the countries.

A review of Export Ratio across countries across the fifteen OECD countries during the years 2000–2010 indicates that Belgium's service economy is the most export-oriented (16.48), followed by Denmark (15.21). The United States is the least export-oriented (2.49) among the nations for which data is provided in this

© Springer International Publishing Switzerland 2015
B. Elango, *Service Industry Databook: Understanding and Analyzing Sector Specific Data Across 15 Nations*, DOI 10.1007/978-3-319-19111-9_4

Table 4.1 Export ratio: Austria

Industry	2000	2001	2002	2003	2004	2005	2006	2007	2008	2009	2010
Telecommunication Services					1.44	2.93	5.08	6.00	5.94	5.46	5.63
Construction					1.48	1.89	2.23	2.22	2.07	2.03	1.66
Air Transportation					12.30	13.53	14.70	15.69	17.35	13.90	16.06
Insurance					0.51	0.72	0.07	0.11	0.13	0.19	0.07
Financial Services					0.17	0.22	0.20	0.26	0.21	0.26	0.23
IT Services	23.07	12.14	9.44	7.23	7.69	12.76	10.44	17.57	17.49	11.89	12.01
Accounting & Legal Services					2.03	2.05	1.82	1.63	1.94	1.84	1.56
Advertising Services					0.82	0.85	0.82	0.82	0.76	0.82	0.69
R&D Services					81.84	86.32	98.67		80.98	64.74	59.22
Engineering Services					2.22	2.25	2.32	2.42	2.25	2.26	2.11
Education					0.94	0.96	0.79	1.05	1.36	1.12	1.13
Personal & Cultural Services					42.60	42.01	36.43	39.10	44.44	44.05	41.69
Health Services					0.09	0.10	0.10	0.08	0.10	0.09	0.10

Note: Export ratio is computed by dividing exports by gross productive output in the corresponding service sector. Exports and gross productive output data are secured from OECD databases.

Table 4.2 Export ratio: Belgium

Industry	2000	2001	2002	2003	2004	2005	2006	2007	2008	2009
Telecommunication Services	14.10	13.62	14.53	14.15	11.48	24.92	26.02	26.38	26.12	29.77
Construction	3.65	5.65	5.41	4.72	4.71	3.07	3.91	3.81	3.53	3.88
Air Transportation	26.99	38.83	88.09	71.08	75.81			89.45	68.00	90.94
Insurance	9.07	13.49	13.72	10.94	10.84	12.63	13.88	12.52	9.95	11.54
Financial Services	12.74	15.80	17.80	18.92	16.64	15.62	13.80	14.10	11.86	13.17
IT Services	30.17	32.11	37.43	36.85	36.59	37.10	39.43	37.31	33.52	40.26
Accounting & Legal Services	6.83	7.34	8.16	8.03	8.47	3.93	6.39	7.71	8.41	8.82
Advertising Services	2.62	3.64	3.53	3.15	2.84	1.82	2.87	3.01	3.59	3.99
R&D Services										
Engineering Services	1.54	1.96	2.13	2.19	2.28	0.48	1.95	1.39	1.50	1.77
Personal & Cultural Services	9.36	12.76	15.60	14.75	13.04	10.52	11.49	11.22	10.61	14.33
All Services	12.88	15.35	17.33	16.01	14.85	17.26	19.83	16.78	16.32	18.22

Note: Export ratio is computed by dividing exports by gross productive output in the corresponding service sector. Exports and gross productive output data are secured from OECD databases.

Table 4.3 Export ratio: Czech Republic

Industry	2000	2001	2002	2003	2004	2005	2006	2007	2008	2009	2010
Telecommunication Services	4.07	7.07	4.42	2.12	4.21	6.68	6.42	7.90	6.55	6.92	7.08
Construction	1.33	1.18	0.72	0.53	0.48	0.86	0.65	0.73	0.90	1.07	2.19
Air Transportation	53.37	52.59	59.14	65.21	75.89	90.41	76.75	68.46	65.07	72.13	73.59
Insurance	0.42	0.73	0.67	0.10	0.37	3.39	2.95	3.21	4.93	6.41	7.92
Financial Services	15.04	7.44	5.82	4.24	8.97	7.10	5.90	3.62	1.71	0.69	0.61
IT Services	9.31	9.93	9.21	3.82	5.79	20.10	23.30	16.42	17.97	17.90	16.72
Accounting & Legal Services	12.77	14.59	8.47	4.54	4.89	9.24	10.45	11.07	12.03	11.45	11.03
Advertising Services	0.34	0.27	0.15	0.10	0.18	0.87	0.82	0.85	0.83	0.94	1.03
R&D Services	6.51	15.81	10.60	2.48	3.53	13.04	15.75	21.42	20.08	18.93	16.60
Engineering Services	0.48	0.43	0.32	0.18	0.14	0.40	0.50	0.68	0.85	0.79	0.85
Education					0.39	0.28	0.07	0.07	0.07	0.08	0.09
Personal & Cultural Services	13.03	10.22	9.28	4.81	7.18	2.94	3.10	4.43	2.23	2.75	3.99
Health Services					0.15	0.14	0.03	0.03	0.03	0.32	0.03
All Services	10.88	10.26	8.32	7.32	7.74	8.39	8.69	8.60	8.52	8.58	9.17

Note: Export ratio is computed by dividing exports by gross productive output in the corresponding service sector. Exports and gross productive output data are secured from OECD databases.

Table 4.4 Export ratio: Denmark

Industry	2000	2001	2002	2003	2004	2005	2006	2007	2008	2009	2010
Telecommunication Services					4.89	7.41	7.49	8.86	7.65	5.71	5.35
Construction						0.65	0.58	0.67	0.59	1.02	1.15
Air Transportation											
Insurance					16.46	1.74	6.16	5.57	7.64	6.05	5.06
Financial Services					3.77	0.99	1.40	1.23	2.51	3.00	3.12
IT Services					12.76	13.83	13.86	12.96	13.27	14.60	17.45
Accounting & Legal Services						5.49	7.06	7.07	7.28	6.48	6.98
Advertising Services						0.59	0.62	0.67	0.70	0.67	1.11
R&D Services								53.47	35.39	28.43	25.03
Engineering Services					2.07	1.34	1.59	1.52	1.71	1.64	1.52
Education			0.02	0.02	0.02	0.02	0.02	0.01		0.04	0.04
Personal & Cultural Services	0.42	0.48	0.41	0.46	0.57	0.51	0.65	0.56	0.57	0.70	0.76
Health Services			0.07	0.05	0.07	0.07	0.09	0.07	0.08	0.08	0.05
All Services	14.22	14.43	13.99	13.50	13.67	15.04	16.63	16.75	17.65	14.94	16.51

Note: Export ratio is computed by dividing exports by gross productive output in the corresponding service sector. Exports and gross productive output data are secured from OECD databases.

Table 4.5 Export ratio: Finland

Industry	2000	2001	2002	2003	2004	2005	2006	2007	2008	2009	2010
Telecommunication Services	3.31	3.46	3.21	2.39	2.78	2.81			3.12	3.04	2.78
Construction	2.35	2.23	1.44	1.96	3.05	2.07	1.15	1.15	3.23	3.83	3.08
Air Transportation	37.88	36.66	39.99	42.87				43.49	44.71	44.05	
Insurance				3.31	3.42	4.10	3.58	5.36	6.67	7.82	4.30
Financial Services					2.95	1.80	1.34	6.42	9.50	7.58	8.32
IT Services	6.98	8.64	13.89	13.15	14.46	26.42	22.38	25.21	99.95	92.61	80.75
Accounting & Legal Services	2.71	4.00	4.18	7.86	6.37	7.41	5.21	4.80	5.15	4.06	3.79
Advertising Services	0.26	0.27	0.37	0.57		0.59	0.36	0.51	0.56	0.80	0.60
R&D Services	24.68	27.13	20.05	27.04		25.08	23.41	23.10	23.56	21.08	19.48
Engineering Services	1.06	0.80	0.90	0.92	0.87	0.90	0.68	0.68	0.70	0.65	0.55
Education	0.04	0.07	0.01		0.01			0.05	0.04	0.01	0.01
Personal & Cultural Services	0.21	0.40	0.48	1.02	0.50	0.36	0.64				
Health Services											
All Services	6.42	7.73	8.17	7.69	8.89	9.28	8.77	10.44	13.16	12.19	11.09

Note: export ratio is computed by dividing exports by gross productive output in the corresponding service sector. Exports and gross productive output data are secured from OECD databases.

Table 4.6 Export ratio: France

Industry	2000	2001	2002	2003	2004	2005	2006	2007	2008	2009	2010
Telecommunication Services	2.86							6.12	5.75	6.31	6.93
Construction	1.76	1.74	1.84	1.42	1.38	1.43	1.44	1.59	1.83	3.31	2.82
Air Transportation	64.17	60.88	60.04	66.23	72.74	70.55	67.43	68.00	67.27	72.63	71.12
Insurance	4.71	3.62	3.82	5.72	2.11	2.46	1.60	1.94	1.48	7.23	6.48
Financial Services	1.49	1.44	1.39	1.24	1.53	1.45	1.31	1.55	1.69	1.75	2.40
IT Services	1.89	2.49	2.65	2.51	2.63	2.83	2.92	2.58	2.25	4.78	4.34
Accounting & Legal Services	3.76							3.53	3.65	5.98	5.85
Advertising Services	0.36	0.39	0.38	0.39	0.44	0.47	0.35	0.27	0.28	1.11	0.87
R&D Services	4.42							7.05	7.39	11.41	11.40
Engineering Services	0.68							0.35	0.35	0.59	0.59
Education								0.19	0.20	0.28	0.34
Personal & Cultural Services	6.25	5.51	5.31	5.49	5.94	5.18	3.86	3.92	4.15	5.03	6.18
Health Services								0.05	0.05	0.07	0.09
All Services	5.28	5.31	5.36	5.31	5.45	5.43	5.33	5.60	5.81	7.07	6.67

Note: Export ratio is computed by dividing exports by gross productive output in the corresponding service sector. Exports and gross productive output data are secured from OECD databases.

Table 4.7 Export ratio: Germany

Industry	2000	2001	2002	2003	2004	2005	2006	2007	2008	2009	2010
Telecommunication Services	2.66	2.61	2.68	3.15	3.65	3.71	4.69	4.62	4.62	3.84	4.20
Construction	1.95	2.19	2.72	3.49	3.32	4.91	5.33	5.30	5.74	5.03	4.12
Air Transportation	43.40	47.73	59.41	55.74	57.56	65.85	63.51	61.89	63.45	70.57	71.26
Insurance	1.09	3.27	13.09	9.51	4.42	2.65	4.38	6.63	5.13	5.93	6.37
Financial Services	3.49	3.24	3.28	3.17	3.75	4.47	5.74	7.76	8.50	7.91	6.87
IT Services	10.14	11.69	13.77	15.23	16.99	16.01	18.17	19.81	21.95	20.32	21.14
Accounting & Legal Services	4.35	5.10	5.18	5.38	5.89	6.19	6.89	8.18	9.18	10.50	10.05
Advertising Services	0.27	0.32	0.35	0.40	0.72	0.76	0.78	0.95	1.03	1.09	1.12
R&D Services	31.21	30.10	32.88	37.29	38.85	49.69	53.65	56.63	56.73	61.50	63.69
Engineering Services	0.67	0.74	0.77	1.59	2.00	1.76	1.74	1.66	1.97	1.99	1.78
Personal & Cultural Services	1.06	1.42	1.41	2.34	2.04	2.43	1.80	1.97	1.79	2.09	1.76
All Services	4.15	4.53	5.12	5.38	5.80	6.14	6.69	7.25	7.81	7.62	7.29

Note: Export ratio is computed by dividing exports by gross productive output in the corresponding service sector. Exports and gross productive output data are secured from OECD databases.

Table 4.8 Export ratio: Hungary

Industry	2000	2001	2002	2003	2004	2005	2006	2007	2008	2009	2010
Telecommunication Services					7.02	6.84	7.88	8.10	9.19	9.05	8.93
Construction	1.87	1.51	2.13	2.00	1.38	1.28	2.85	3.47	3.62	3.43	2.94
Air Transportation											
Insurance	2.51	2.41	1.49	2.90	2.41	0.44	0.84	1.20	1.28	0.93	1.74
Financial Services	7.70	8.03	5.33	5.13	5.57	2.69	2.75	3.69	3.39	2.68	2.47
IT Services	15.14	18.00	15.97	15.16	17.62	18.50	21.78	27.77	31.45	34.06	33.80
Accounting & Legal Services					9.85	14.67	16.03	19.12	18.61	20.22	22.56
Advertising Services					1.00	0.98	1.17	1.39	1.53	1.65	1.98
R&D Services					34.31	30.37	38.25	49.75	45.65	44.69	58.83
Engineering Services					0.44	0.72	0.77	1.19	1.09	1.41	1.39
Education					0.27	0.30	0.21	0.15	0.08	0.09	0.06
Personal & Cultural Services	19.32	49.56	31.24	43.93	55.06	53.96	45.27	48.31	30.69	40.79	46.26
Health Services					0.03	0.04	0.07	0.12	0.16	0.38	0.52
All Services	13.79	13.04	10.71	11.32	11.09	11.89	12.13	12.77	12.86	14.14	15.01

Note: Export ratio is computed by dividing exports by gross productive output in the corresponding service sector. Exports and gross productive output data are secured from OECD databases.

Table 4.9 Export ratio: Italy

Industry	2000	2001	2002	2003	2004	2005	2006	2007	2008	2009	2010
Telecommunication Services	3.11	3.25	2.22	3.87	3.36	3.56	5.29	4.80	10.02	8.18	10.91
Construction	0.89	1.08	1.05	1.07	0.90	0.90	0.93	1.09	0.10	0.14	0.04
Air Transportation	30.79	28.05	37.58	32.02	52.84	55.49	57.45	60.81	52.41	53.46	50.72
Insurance	7.13	7.45	8.46	5.69	7.83	6.37	6.81	4.81	13.51	14.92	11.85
Financial Services	0.69	0.70	1.10	1.27	1.17	1.47	2.14	3.70	2.68	1.95	2.25
IT Services	1.46	1.07	1.03	1.18	1.26	1.30	1.81	1.62	4.02	3.35	3.45
Accounting & Legal Services	1.53	1.82	1.86	1.88	1.90	1.94	2.22	2.41	1.79	1.45	2.31
Advertising Services	0.30	0.30	0.29	0.26	0.29	0.28	0.27	0.29	0.41	0.36	0.40
R&D Services	6.36	9.40	10.11	8.06	8.51	7.34	8.94	9.00	16.16	16.55	14.01
Engineering Services	0.91	0.82	0.84	0.81	0.77	0.90	1.05	1.13	0.93	0.46	0.78
Education		0.08	0.23	0.19	0.11	0.11	0.12	0.18	0.03	0.04	0.03
Personal & Cultural Services	2.85	2.80	3.38	2.87	2.60	2.52	2.97	3.20	0.71	0.91	0.92
Health Services		0.03	0.10	0.08	0.04	0.04	0.04	0.06	0.03	0.03	0.03
All Services	4.71	4.79	4.71	4.82	5.05	5.04	5.27	5.45	5.31	4.60	4.55

Note: Export ratio is computed by dividing exports by gross productive output in the corresponding service sector. Exports and gross productive output data are secured from OECD databases.

Table 4.10 Export ratio: Korea

Industry	2000	2001	2002	2003	2004	2005	2006	2007	2008	2009	2010
Telecommunication Services							0.97	1.02	1.32	1.52	1.47
Construction	1.13	1.52	2.46	1.87	2.17	3.37	4.52	5.60	8.65	10.26	7.59
IT Services	0.39	0.54	0.56	0.76	0.50	0.90	3.19	3.94	3.79	2.53	2.19
Advertising Services							0.41	0.41	0.46	0.40	0.36
R&D Services							5.17	8.14	9.30	9.11	5.27
Education							0.03	0.06	0.10	0.09	0.08
Personal & Cultural Services	1.57	1.52	1.59	0.59	0.90	1.58	1.88	1.98	2.56	2.82	2.86
Health Services			0.02	0.02	0.03	0.02	0.03	0.02	0.03	0.03	0.02
All Services	6.83	6.73	5.90	6.06	6.98	6.56	6.53	7.48	10.09	9.19	9.15

Note: Export ratio is computed by dividing exports by gross productive output in the corresponding service sector. Exports and gross productive output data are secured from OECD databases.

Table 4.11 Export ratio: Netherlands

Industry	2000	2001	2002	2003	2004	2005	2006	2007	2008	2009	2010
Telecommunication Services		9.52	6.97	6.68	7.62	8.39	8.41	8.14			
Construction	4.37	5.33	4.27	2.73	2.89	3.51	2.65	2.56	2.93	2.87	2.92
Air Transportation		86.72	74.52	95.17		92.84	74.04	75.38	78.27	75.78	77.64
Insurance	1.57	1.72	2.26	1.88	2.08	1.99	2.07	2.37	2.94	2.45	2.67
Financial Services	2.73	3.00	2.79	2.14	2.40	2.48	3.57	4.16	3.29	2.47	1.93
IT Services	9.33	6.58	11.28	22.58	23.78	21.40	25.08	27.40	25.74	25.73	25.88
Accounting & Legal Services	9.79	19.68	28.59	6.90	5.47	5.77	7.40	8.41	8.76	8.64	8.19
Advertising Services	1.29			0.31	0.24	0.21	0.18	0.16	0.14	0.43	0.46
R&D Services	46.01			91.35	87.12		84.78	84.55	78.46	66.58	63.19
Engineering Services	2.06	3.44	3.71	0.95	1.03	0.73	0.60	0.54	0.52	0.64	0.65
Education			0.01	0.01	0.01	0.01	0.01	0.01		0.02	0.02
Personal & Cultural Services	7.51	6.94	6.47	5.86	7.18	7.98	5.88	5.01	5.25	5.72	4.85
Health Services			0.03	0.02	0.03	0.03	0.04	0.03	0.04	0.04	0.02
All Services	11.48	11.10	11.44	12.59	13.51	13.74	13.57	14.09	14.65	13.78	13.56

Note: Export ratio is computed by dividing exports by gross productive output in the corresponding service sector. Exports and gross productive output data are secured from OECD databases.

Table 4.12 Export ratio: Norway

Industry	2000	2001	2002	2003	2004	2005	2006	2007	2008	2009	2010
Telecommunication Services	3.53	4.02	3.67		3.01	3.48	3.40	4.30	5.23	4.95	0.07
Construction	0.41	0.49			0.38	0.70	0.77	0.61	0.73	0.83	
Air Transportation	17.79	10.73	8.77	9.11	12.56	13.51	10.18	12.43	15.21	6.04	
Insurance	39.61	12.16	18.63	14.67	10.10	8.72	7.49	5.61	4.00	5.31	
Financial Services	3.57	5.05	5.44		6.15	6.39	7.13	7.37	9.43	9.55	
IT Services	9.03	14.91		8.47	11.03	14.93	20.45	13.82	22.29	29.65	
Accounting & Legal Services	6.75	5.23		4.63	4.38	6.10	6.83	4.51	5.92	7.29	
Advertising Services	0.61	0.28			0.29	0.40	0.71	0.49	0.43	0.68	
R&D Services	15.34	17.71		14.67	14.35	23.57	35.00	25.35	26.58	34.07	
Engineering Services	1.71	1.76	1.53	1.55	1.51	1.87	2.22	1.75	1.72	1.74	
Education			0.02	0.02	0.02	0.02	0.02	0.01		0.04	
Personal & Cultural Services	0.44	0.53	0.39	0.45	0.56	0.49	0.61	0.57	0.59	0.73	
Health Services			0.07	0.05	0.08	0.07	0.09	0.07	0.08	0.09	
All Services	11.41	11.02	10.05	10.02	10.53	10.87	11.33	11.44	11.21	10.68	

Note: Export ratio is computed by dividing exports by gross productive output in the corresponding service sector. Exports and gross productive output data are secured from OECD databases.

Table 4.13 Export ratio: Slovenia

Industry	2000	2001	2002	2003	2004	2005	2006	2007	2008	2009	2010
Telecommunication Services								9.36	17.34	16.47	16.98
Construction								1.93	3.52	2.73	2.26
Air Transportation								74.93	96.67	99.86	87.74
Insurance								2.72	14.12	10.32	9.21
Financial Services								2.80	1.65	1.64	1.70
IT Services								16.97	18.58	15.49	14.00
Accounting & Legal Services								10.28	9.55	6.60	5.82
Advertising Services								1.99	2.02	2.08	2.01
R&D Services								22.80	26.80	11.77	7.95
Engineering Services								0.36	0.39	0.46	0.36
Education								0.11	0.03	0.30	0.30
Personal & Cultural Services								3.44	3.45	4.24	4.47
Health Services								0.65	0.72	0.72	0.40
All Services								13.54	15.12	13.67	13.08

Note: Export ratio is computed by dividing exports by gross productive output in the corresponding service sector. Exports and gross productive output data are secured from OECD databases.

Table 4.14 Export ratio: Sweden

Industry	2000	2001	2002	2003	2004	2005	2006	2007	2008	2009	2010
Telecommunication Services	6.86	8.14	6.17	5.72	8.97	10.23	10.58	10.59	12.34	11.81	10.52
Construction	3.10	2.87	2.06	2.49	2.21	1.86	1.62	1.95	1.99	1.63	1.76
Air Transportation	25.54	29.85	34.85	54.67	58.22	49.73	49.76	56.53	56.32	62.23	82.47
Insurance	11.43	11.69	11.72	14.34	12.10	13.14	12.98	11.35	11.15	12.02	11.42
Financial Services	6.94	6.45	5.78	8.07	9.13	10.03	12.41	8.74	8.78	6.95	6.01
IT Services	11.75	14.33	15.32	16.80	16.74	16.59	21.74	33.33	36.51	40.77	36.58
Accounting & Legal Services	7.96	7.77	7.05	9.20	9.18	10.65	11.22	10.97	9.96	9.36	11.28
Advertising Services	0.65	0.73	0.70	0.74	0.68	0.79	0.73	0.78	0.77	0.93	1.07
Education							0.22	0.19	0.44	0.37	0.35
Personal & Cultural Services	2.34	2.49	2.17	3.12	2.34	2.83	3.08	3.24	4.73	3.90	3.93
Health Services							0.02	0.02	0.04	0.03	0.02
All Services	7.90	8.91	8.57	8.84	9.72	10.35	10.84	12.00	12.44	12.10	11.93

Note: Export ratio is computed by dividing exports by gross productive output in the corresponding service sector. Exports and gross productive output data are secured from OECD databases.

Table 4.15 Export ratio: United States

Industry	2000	2001	2002	2003	2004	2005	2006	2007	2008	2009	2010
Telecommunication Services	0.77	0.83	0.75	0.85	0.86	0.83	1.20	1.33	1.57	1.61	1.72
Construction	0.19	0.24	0.29	0.21	0.18	0.11	0.14	0.21	0.31	0.37	0.27
Air Transportation	20.63	20.68	21.74	18.63	20.96	22.99	22.58	24.24	27.62	27.67	29.28
Insurance	0.84	0.76	0.94	1.17	1.30	1.26	1.51	1.64	2.02	2.34	2.33
Financial Services	3.94	3.57	3.62	3.82	4.79	4.78	5.38	6.59	6.44	6.33	6.82
IT Services	2.62	2.55	2.69	3.10	3.14	3.18	3.12	3.32	3.39	3.59	3.39
Education	1.57	1.63	1.73	1.72	1.68	1.63	1.60	1.64	1.75	1.91	1.99
Personal & Cultural Services	0.10	0.11	0.10	0.11	0.12	0.13	0.21	0.28	0.33	0.34	0.45
All Services	2.34	2.18	2.16	2.12	2.30	2.35	2.47	2.74	2.91	2.85	2.96

Note: Export ratio is computed by dividing exports by gross productive output in the corresponding service sector. Exports and gross productive output data are secured from OECD databases.

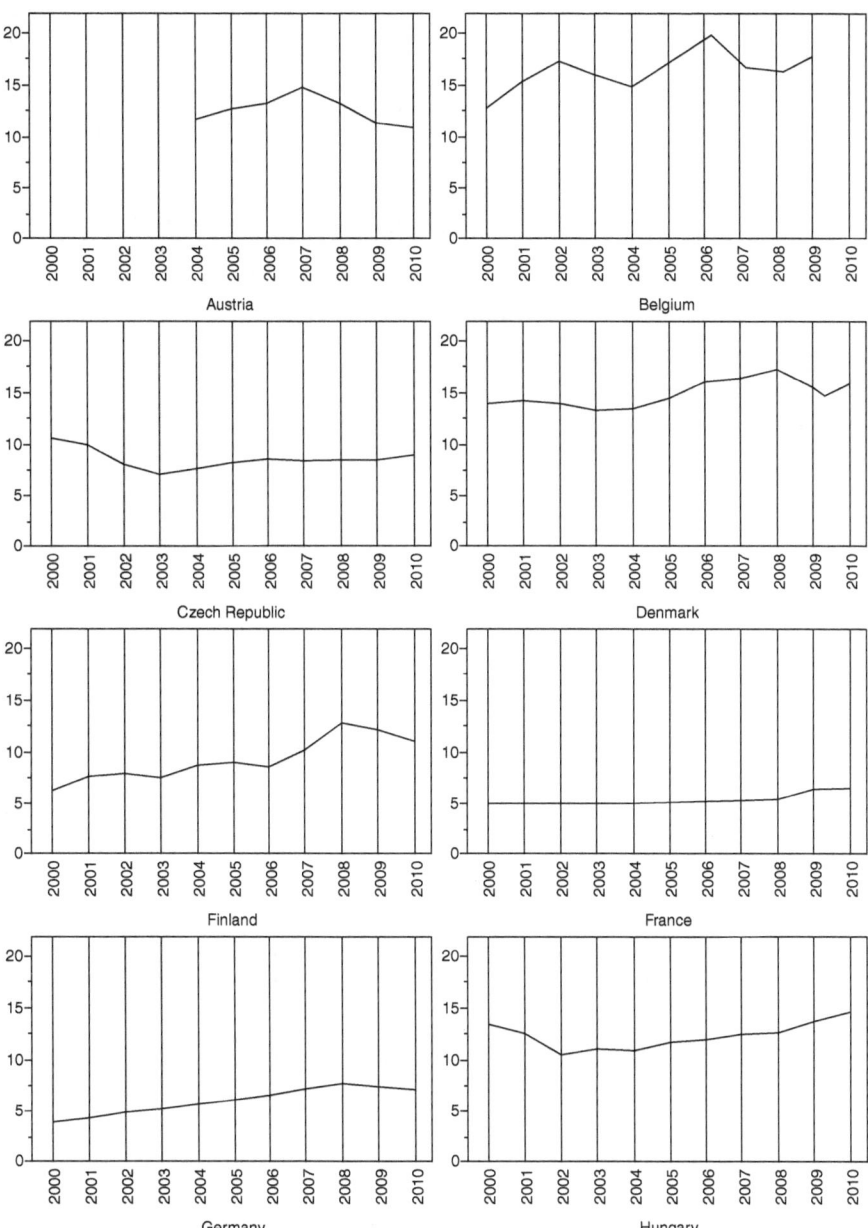

Fig. 4.1 Export ratio for service sector across countries (2000–2010)

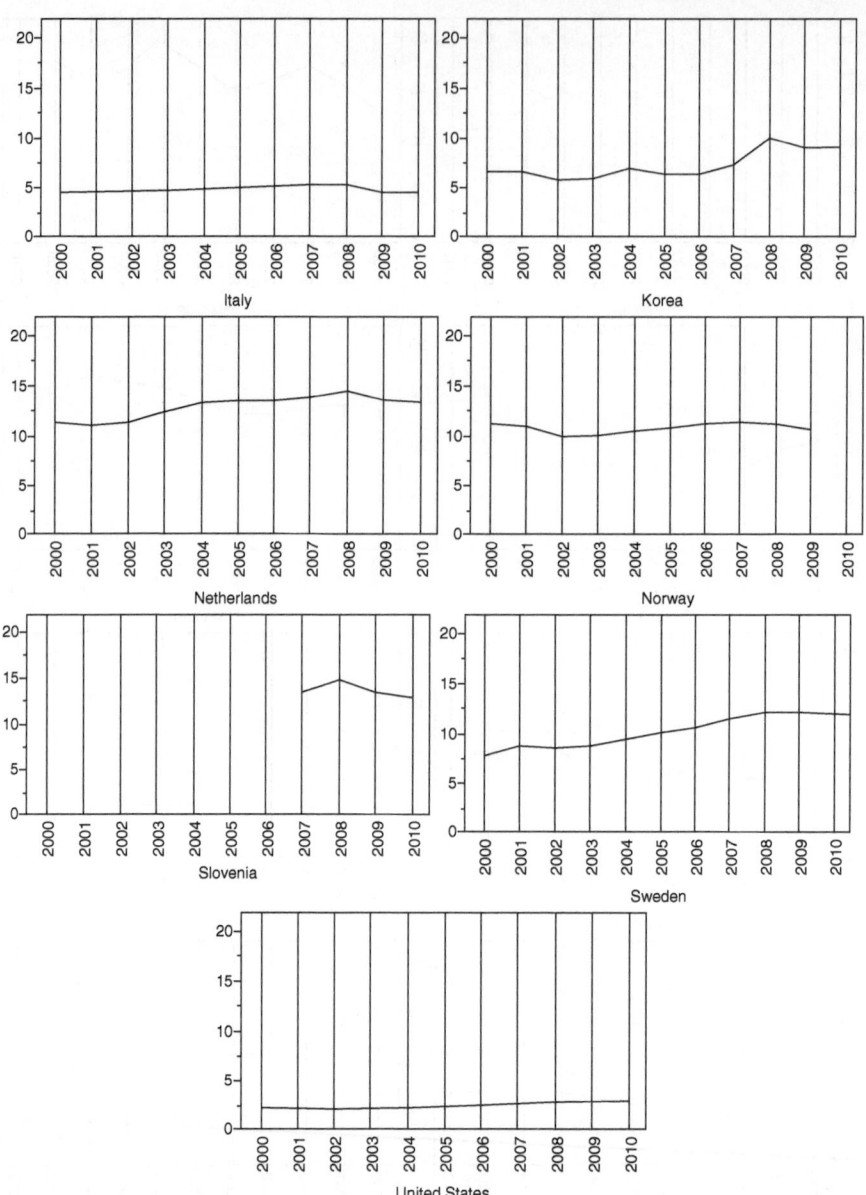

Fig. 4.1 (continued)

book, followed by Italy (4.94). Overall, a review of Export Ratio indicates these numbers have an increasing trend for the majority of the countries.

A word of caution is needed when interpreting these numbers, as the context is critical. Export Ratio captures the extent to which a country's products in an industry are exported, but it does not capture the actual volume of exports from a specific industry in a country. A failure to recognize this difference could create a misleading picture if all facts are not taken into account. For instance, as noted earlier, Belgium exports a greater portion of the exports compared to the United States, but the United States is a far greater exporter of services in aggregate value terms than Belgium. The obvious reason for the United States having a lower Export Ratio number is the fact that the U.S. economy is one of the largest in the world and is dominant in most service sectors [see the discussion on Domestic Global Ratio (Section 4.4) in this regard]. Therefore the reader is advised to review the aggregate numbers to understand the context of the ratios presented. In Chapter 2 of this book, the aggregate numbers for exports are presented in the case of the United States, but for other countries, the reader should refer to another resource, such as the OECD databases.

4.2 Import Ratio

Import Ratio represents the ratio of the value of services that have been transferred from a foreign country relative to the level of economic output that takes place in a particular industry in a country. Stated differently, it represents the extent of market share held by imports in an industry and indicates the extent to which a market is controlled by non-domiciled foreign firms. Import Ratio indicates the diversity (local vs. foreign) in competition and could play a significant role in impacting profitability for firms. Import Ratio numbers were derived from OECD databases as follows: Initially, imports in a particular service sector and the gross productive output were determined from the OECD database for each service industry segment in a country. Imports from the country was then divided by the gross productive output in the corresponding service sector and multiplied by 100 to arrive at the Import Ratio.

Import Ratio numbers for the years 2000–2010 for 15 OECD countries across 11–12 industries are presented where data was available. Table 4.16 shows the Import Ratio for Austria; Table 4.17 shows the Import Ratio for Belgium; Table 4.18 shows the Import Ratio for the Czech Republic; Table 4.19 shows the Import Ratio for Denmark; Table 4.20 shows the Import Ratio for Finland; Table 4.21 shows the Import Ratio for France; Table 4.22 shows the Import Ratio for Germany; Table 4.23 shows the Import Ratio for Hungary; Table 4.24 shows the Import Ratio for Italy; Table 4.25 shows the Import Ratio for Korea; Table 4.26 shows the Import Ratio for the Netherlands; Table 4.27 shows the Import Ratio for Norway; Table 4.28 shows the Import Ratio for Slovenia; Table 4.29 shows the Import Ratio for Sweden; and Table 4.30 shows the Import Ratio for the United States. Figure 4.2 graphically shows the average Import Ratio for services sector in each of the countries.

Table 4.16 Import ratio: Austria

Industry	2000	2001	2002	2003	2004	2005	2006	2007	2008	2009	2010
Telecommunication Services					6.66	7.39	9.82	11.11	10.98	10.70	9.69
Construction					2.66	2.12	1.77	2.40	2.64	2.04	1.62
Air Transportation					71.50	72.86	76.69	73.01	80.47	71.64	72.28
Insurance					10.73	13.10	15.74	22.47	13.48	17.29	12.53
Financial Services	8.42	5.79	5.08	3.35	2.48	5.06	3.57	3.78	3.57	2.27	2.21
IT Services					12.67	13.12	14.08	17.09	19.44	18.11	16.88
Accounting & Legal Services					6.16	6.28	6.50	6.20	5.99	5.42	4.79
Advertising Services					13.57	13.43	14.56	14.12	16.20	15.56	15.07
R&D Services					31.39	33.68	41.65	43.15	36.50	38.09	27.43
Engineering Services					0.73	0.96	1.07	1.16	1.24	1.24	1.18
Education					0.32	0.31	0.30	0.36	0.41	0.31	0.28
Personal & Cultural Services					16.93	15.93	16.51	15.44	15.32	14.60	15.77
Health Services					2.62	2.57	2.76	2.59	2.65	2.57	2.61
All Services	8.37	9.01	9.09	9.94	10.38	10.46	10.51	11.07	11.22	10.26	9.67

Note: Import ratio is computed by dividing imports by gross productive output in the corresponding service sector. Imports and gross productive output data are secured from OECD databases.

Table 4.17 Import ratio: Belgium

Industry	2000	2001	2002	2003	2004	2005	2006	2007	2008	2009	2010
Telecommunication Services					3.77	5.70	5.27	5.85	5.75	4.56	4.30
Construction						0.38	0.31	0.24	0.36	1.05	0.95
Air Transportation					30.98	53.85	52.48	54.15	56.65	94.38	83.15
Insurance					6.52	6.68	6.48	5.84	2.21	5.47	5.99
Financial Services					3.14	0.58	0.71	0.83	1.19	1.25	1.20
IT Services					10.51	14.25	14.72	14.87	16.69	13.86	15.39
Accounting & Legal Services					9.77	1.53	1.97	2.59	3.05	1.99	1.69
Advertising Services						5.32	6.34	7.39	8.00	7.66	6.66
R&D Services						28.52	32.33	40.17	41.77	35.15	42.51
Engineering Services					0.77	1.03	0.93	1.17	1.41	0.87	0.77
Education			0.05	0.06	0.07	0.09	0.10	0.15	0.02	0.03	0.03
Personal & Cultural Services	1.17	1.28	1.31	1.42	1.37	1.33	1.50	1.73	0.50	0.78	0.85
Health Services			0.02	0.02	0.02	0.02	0.03	0.03	0.03	0.04	0.04
All Services	7.21	7.56	7.98	7.93	8.30	8.53	9.74	10.49	11.11	9.96	9.34

Note: Import ratio is computed by dividing imports by gross productive output in the corresponding service sector. Imports and gross productive output data are secured from OECD databases.

Table 4.18 Import ratio: Czech Republic

Industry	2000	2001	2002	2003	2004	2005	2006	2007	2008	2009	2010
Telecommunication Services					9.17	12.97	11.05	10.92	9.51	8.32	8.45
Construction						0.70	0.55	0.36	0.51	1.68	1.67
Air Transportation					79.67						
Insurance					26.62	30.56	22.99	18.95	5.67	15.67	19.86
Financial Services					14.66	2.28	2.75	2.80	3.26	3.80	3.76
IT Services					33.87	35.26	32.75	27.60	25.79	22.74	26.23
Accounting & Legal Services						14.90	18.79	21.28	20.06	13.21	11.90
Advertising Services						9.21	10.76	10.81	9.84	10.78	10.09
R&D Services						90.60	96.47		90.88	84.13	96.89
Engineering Services					1.30	1.57	1.39	1.41	1.49	1.00	0.96
Education			0.22	0.21	0.27	0.30	0.34	0.45	0.06	0.08	0.08
Personal & Cultural Services					10.56	37.65	31.85	37.68	32.18	22.96	22.12
Health Services			0.10	0.08	0.10	0.10	0.13	0.10	0.09	0.11	0.14
All Services	33.59	32.29	28.91	26.21	26.74	26.21	27.69	26.85	24.39	23.22	23.39

Note: Import ratio is computed by dividing imports by gross productive output in the corresponding service sector. Imports and gross productive output data are secured from OECD databases.

Table 4.19 Import ratio: Denmark

Industry	2000	2001	2002	2003	2004	2005	2006	2007	2008	2009	2010
Telecommunication Services					5.51	8.36	7.77	8.74	8.39	6.85	7.11
Construction						0.62	0.52	0.39	0.64	2.15	2.32
Air Transportation					37.35	61.54	58.41	54.32	66.74		76.78
Insurance					12.40	12.05	10.94	10.69	4.12	9.81	13.51
Financial Services					5.46	1.00	1.30	1.34	1.63	1.84	1.95
IT Services					11.42	14.90	15.94	15.66	16.79	14.18	16.85
Accounting & Legal Services					54.62	8.88	10.82	13.51	14.96	9.90	8.97
Advertising Services						10.15	11.45	12.11	12.70	12.11	12.16
R&D Services						35.92	38.65	43.16	27.73	25.26	28.91
Engineering Services					0.73	0.89	0.84	0.95	1.16	0.82	0.91
Education			0.08	0.08	0.09	0.12	0.14	0.21	0.03	0.04	0.04
Personal & Cultural Services	1.10	1.09	1.03	1.11	1.12	1.10	1.29	1.48	0.42	0.58	0.69
Health Services			0.04	0.03	0.04	0.04	0.06	0.05	0.05	0.06	0.07
All Services	12.51	12.69	12.80	11.98	12.43	12.82	14.38	14.72	15.17	13.91	14.19

Note: Import ratio is computed by dividing imports by gross productive output in the corresponding service sector. Imports and gross productive output data are secured from OECD databases.

Table 4.20 Import ratio: Finland

Industry	2000	2001	2002	2003	2004	2005	2006	2007	2008	2009	2010
Telecommunication Services	5.32	5.73	3.92	3.22	3.05	6.64			7.53	8.02	7.05
Construction	0.28	0.63	1.16	1.41	2.52	1.75	1.09	0.92	1.48	1.52	1.40
Air Transportation	28.14	39.35	27.17	29.95				31.09	42.22	38.79	
Insurance	1.60	3.56	4.06	4.89	8.00	9.62	10.27	13.56	14.00	14.38	15.73
Financial Services					1.54	2.12	3.14	3.32	2.95	2.99	8.15
IT Services	10.40	9.74	10.29	11.21	14.09	20.23	16.94	20.60	23.32	44.38	26.78
Accounting & Legal Services	7.60	6.13	9.41	10.06	10.57	11.96	8.34	12.39	10.97	8.66	8.03
Advertising Services	22.14	20.74	33.70	38.35		48.92	46.66	59.13	86.65	85.91	81.41
R&D Services	94.02										
Engineering Services	0.31	0.31	0.25	0.26	0.29	0.37	0.38	0.31	0.34	0.27	0.28
Education	0.09	0.11	0.05	0.12	0.14	0.07					
Personal & Cultural Services	2.66	1.70	3.41	1.73	0.89	1.48	1.47				
All Services	7.86	7.76	7.74	8.16	8.54	9.69	9.41	10.15	12.74	11.96	10.94

Note: Import ratio is computed by dividing imports by gross productive output in the corresponding service sector. Imports and gross productive output data are secured from OECD databases.

Table 4.21 Import ratio: France

Industry	2000	2001	2002	2003	2004	2005	2006	2007	2008	2009	2010
Telecommunication Services	2.58							3.70	3.94	5.94	5.51
Construction	0.94	0.99	0.95	0.68	0.76	0.64	0.62	0.64	0.71	1.80	1.96
Air Transportation	60.08	59.51	55.01	64.36	79.65	75.74	69.36	67.61	63.85	63.72	76.35
Insurance		2.56	5.43	6.48	3.34	4.91	4.60	3.85	3.61	5.55	5.42
Financial Services	1.73	1.93	1.93	2.21	2.64	2.41	2.03	1.67	1.65	1.46	1.72
IT Services	1.75	2.16	2.67	2.47	2.56	2.95	2.96	3.09	2.75	6.66	5.03
Accounting & Legal Services	3.46							3.00	3.13	4.24	4.23
Advertising Services	4.87	4.74	5.43	5.50	6.62	7.07	5.93	5.35	7.66	13.07	11.52
R&D Services	2.89							10.60	11.21	14.33	14.62
Engineering Services	0.59							0.57	0.57	0.81	0.82
Education								0.20	0.22	0.20	0.20
Personal & Cultural Services	7.72	7.93	7.27	6.87	6.83	6.80	5.94	6.27	6.84	6.05	5.97
Health Services								0.07	0.08	0.07	0.07
All Services	3.99	4.13	4.29	4.45	4.74	4.75	4.69	4.86	4.96	6.13	5.94

Note: Import ratio is computed by dividing imports by gross productive output in the corresponding service sector. Imports and gross productive output data are secured from OECD databases.

Table 4.22 Import ratio: Germany

Industry	2000	2001	2002	2003	2004	2005	2006	2007	2008	2009	2010
Telecommunication Services	5.80	5.38	5.44	4.79	4.79	5.15	6.45	6.38	6.15	5.31	6.27
Construction	2.26	2.46	2.46	2.67	2.75	3.11	3.17	3.59	4.06	3.50	2.57
Air Transportation	34.97	38.75	51.70	52.49	51.55	59.40	60.28	60.34	66.95	64.10	67.06
Insurance	1.71	2.17	2.41	4.96	6.39	5.77	3.16	3.73	4.75	4.00	4.31
Financial Services	1.98	2.27	2.01	1.90	2.57	3.10	3.87	5.25	4.88	4.28	3.87
IT Services	13.27	14.75	15.42	16.55	17.12	16.34	16.85	18.54	19.50	17.18	18.18
Accounting & Legal Services	9.67	10.76	10.00	9.85	10.31	9.80	10.29	11.45	11.57	14.02	14.28
Advertising Services	8.93	10.43	9.03	8.94	9.93	10.55	10.46	10.87	13.28	12.54	12.88
R&D Services	30.86	42.39	42.29	34.96	33.01	38.06	37.63	40.17	37.98	49.35	48.84
Engineering Services	0.94	1.24	1.27	1.35	1.48	1.46	1.37	1.34	1.35	1.48	1.35
Personal & Cultural Services	9.48	10.89	6.00	6.87	6.69	7.17	8.99	4.91	4.79	4.72	4.42
All Services	6.90	7.30	7.28	7.59	7.82	7.97	8.06	8.53	8.95	8.36	8.02

Note: import ratio is computed by dividing imports by gross productive output in the corresponding service sector. Imports and gross productive output data are secured from OECD databases.

Table 4.23 Import ratio: Hungary

Industry	2000	2001	2002	2003	2004	2005	2006	2007	2008	2009	2010
Telecommunication Services					7.18	7.13	9.61	10.18	9.13	9.90	9.53
Construction	1.11	1.35	2.90	2.05	0.51	0.76	1.72	2.33	2.44	2.78	2.59
Air Transportation											
Insurance	13.31	17.05	16.60	18.61	13.55	11.60	12.16	11.12	8.92	13.67	9.07
Financial Services	11.73	10.57	5.44	7.44	8.39	3.42	3.61	4.22	4.22	3.82	3.28
IT Services	15.85	16.41	13.59	16.53	20.56	23.86	24.03	22.80	22.54	20.54	20.18
Accounting & Legal Services					11.66	15.30	16.98	18.24	16.73	19.43	20.10
Advertising Services					6.69	6.93	8.81	9.47	11.32	10.61	10.92
R&D Services					95.09	94.64	65.81	81.31	94.28	87.56	
Engineering Services					0.58	0.79	1.23	1.25	1.33	1.20	1.08
Education					0.40	0.37	0.33	0.51	0.38	0.33	0.33
Personal & Cultural Services	14.24	46.53	29.14	43.90	49.44	48.48	36.54	34.58	28.17	34.18	35.41
Health Services					0.06	0.01	0.11	0.20	0.20	0.36	0.41
All Services	11.22	10.21	9.91	11.25	10.47	10.50	10.52	11.41	11.48	12.04	12.00

Note: Import ratio is computed by dividing imports by gross productive output in the corresponding service sector. Imports and gross productive output data are secured from OECD databases.

Table 4.24 Import ratio: Italy

Industry	2000	2001	2002	2003	2004	2005	2006	2007	2008	2009	2010
Telecommunication Services	4.57	4.28	3.74	5.15	3.94	4.39	7.19	6.95	10.59	8.54	10.45
Construction	0.75	1.06	1.38	1.25	1.19	0.89	0.98	1.20	0.02	0.01	0.03
Air Transportation	46.08	40.56	56.90	56.38	65.56	69.21	65.42	79.41	93.65		
Insurance	9.62	9.45	11.36	9.07	11.40	8.64	11.91	11.37	20.47	18.30	17.58
Financial Services	0.89	0.97	1.05	1.14	1.62	1.44	1.11	1.37	3.39	3.81	4.00
IT Services	3.01	2.91	2.83	2.50	2.70	3.27	3.52	3.27	7.97	6.37	7.44
Accounting & Legal Services	1.93	2.15	2.10	2.38	1.44	1.78	1.66	1.85	3.63	3.24	3.75
Advertising Services	3.42	3.13	2.79	2.81	3.07	3.19	2.98	3.43	7.27	4.33	4.83
R&D Services	5.07	4.63	4.62	4.60	5.14	4.51	4.92	5.83	9.76	12.09	11.04
Engineering Services	1.13	0.96	0.77	0.79	0.73	0.88	0.66	0.78	0.83	0.73	0.50
Education		0.14	0.12	0.11	0.14	0.12	0.13	0.14	0.07	0.10	0.10
Personal & Cultural Services	6.14	5.64	4.42	4.09	5.31	5.70	5.27	5.97	1.98	1.89	1.80
Health Services		0.06	0.05	0.04	0.05	0.04	0.04	0.04	0.03	0.02	0.12
All Services	4.63	4.79	4.94	5.00	4.97	5.07	5.35	5.92	5.89	5.18	5.11

Note: Import ratio is computed by dividing imports by gross productive output in the corresponding service sector. Imports and gross productive output data are secured from OECD databases.

Table 4.25 Import ratio: Korea

Industry	2000	2001	2002	2003	2004	2005	2006	2007	2008	2009	2010
Telecommunication Services							1.41	1.44	1.87	2.43	2.18
Construction	0.23	0.29	0.49	0.36	0.40	0.63	0.84	1.06	1.65	1.98	1.46
IT Services	3.40	3.49	3.59	3.42	3.11	2.87	7.69	6.31	7.13	4.66	4.66
Advertising Services							28.88	29.99	51.31	69.91	58.55
R&D Services							16.43	16.78	18.15	17.93	16.36
Education							0.17	0.21	0.26	0.28	0.31
Personal & Cultural Services	1.83	2.26	2.44	2.04	2.65	2.82	3.42	4.10	4.32	4.56	4.59
Health Services			0.01	0.01	0.02	0.01	0.02	0.02	0.02	0.02	0.02
All Services	7.27	7.40	7.14	7.06	7.91	7.87	8.06	8.71	10.73	10.02	10.05

Note: Import ratio is computed by dividing imports by gross productive output in the corresponding service sector. Imports and gross productive output data are secured from OECD databases.

Table 4.26 Import ratio: Netherlands

Industry	2000	2001	2002	2003	2004	2005	2006	2007	2008	2009	2010
Telecommunication Services		8.76	6.91	7.01	6.77	7.41	8.41	7.61			
Construction	1.65	1.90	1.48	1.54	1.45	1.80	1.38	1.37	1.59	2.09	2.11
Air Transportation	54.09	27.78	28.11	60.51	57.80	52.30	36.10	34.54	35.29	40.50	40.13
Insurance	4.25	4.03	8.09	3.71	3.82	3.55	3.59	3.99	5.01	4.33	4.54
Financial Services	3.57	4.59	4.33	2.78	3.03	3.66	4.77	5.69	4.28	2.57	2.18
IT Services	9.53	10.01	12.59	18.65	20.00	21.30	22.50	23.30	22.21	24.17	21.89
Accounting & Legal Services	11.14	25.30	30.04	5.87	5.22	5.24	5.16	5.38	5.47	5.15	5.19
Advertising Services	13.21			7.83	7.05	7.82	6.02	6.17	6.46	6.98	7.25
R&D Services	34.08						68.36	59.96	56.42	53.95	49.95
Engineering Services	2.19	4.35	4.51	0.82	1.07	0.42	0.31	0.29	0.29	0.33	0.27
Education			0.04	0.04	0.05	0.06	0.07	0.11	0.02	0.02	0.02
Personal & Cultural Services	8.11	7.50	7.47	7.60	8.17	8.34	6.00	5.32	5.03	5.73	4.87
Health Services			0.02	0.01	0.02	0.02	0.03	0.02	0.02	0.03	0.03
Total Services	11.67	11.64	11.65	12.55	12.69	12.62	12.19	12.40	13.04	13.12	12.21

Note: Import ratio is computed by dividing imports by gross productive output in the corresponding service sector. Imports and gross productive output data are secured from OECD databases.

Table 4.27 Import ratio: Norway

Industry	2000	2001	2002	2003	2004	2005	2006	2007	2008	2009	2010
Telecommunication Services	2.57	3.28	2.99		2.34	2.53	2.42	2.93	8.95	7.85	
Construction	0.16	0.13		0.15	0.13	0.11	0.10	0.10	0.11	0.10	
Air Transportation	32.24	34.81	34.40	30.11	30.01	21.68	31.00	40.29	15.71	13.65	
Insurance	49.41	33.72	29.92	22.44	8.39	10.68	4.82	4.25	1.84	2.44	
Financial Services	1.83	2.28	2.64	2.64	2.32	10.78	7.64	8.09	7.41	7.10	
IT Services	6.84	9.47	14.27	11.51	12.23	18.87	18.82	20.04	17.67	17.68	
Accounting & Legal Services	4.32	3.36	2.43	1.94	2.90	5.31	4.35	4.39	4.75	4.76	
Advertising Services	4.89	2.90	1.87		2.00	3.60	4.72	4.97	3.62	4.60	
R&D Services	15.20	13.86			9.31	20.16	21.87	24.25	20.95	21.84	
Engineering Services	0.97	0.52	0.39	0.37	0.44	0.69	0.60	0.43	0.46	0.45	
Education			0.08	0.08	0.11	0.13	0.15	0.21	0.03	0.04	
Personal & Cultural Services	1.16	1.20	0.99	1.07	1.09	1.04	1.21	1.50	0.44	0.61	
Health Services			0.04	0.03	0.04	0.04	0.06	0.05	0.05	0.06	
All Services	10.15	9.52	9.06	9.23	9.34	10.96	10.85	11.19	11.14	10.20	

Note: Import ratio is computed by dividing imports by gross productive output in the corresponding service sector. Imports and gross productive output data are secured from OECD databases.

Table 4.28 Import ratio: Slovenia

Industry	2000	2001	2002	2003	2004	2005	2006	2007	2008	2009	2010
Telecommunication Services								11.54	18.25	18.67	18.58
Construction								2.05	1.82	1.37	1.03
Air Transportation								58.91	91.93	87.00	77.13
Insurance								3.43	15.86	12.09	11.07
Financial Services								3.93	1.75	2.43	1.86
IT Services								21.12	17.32	19.99	16.54
Accounting & Legal Services								13.32	10.36	9.32	7.28
Advertising Services								28.15	28.85	24.85	29.65
R&D Services								28.65	25.22	19.29	14.54
Engineering Services								0.33	0.34	0.47	0.38
Education								1.59	0.22	0.29	0.28
Personal & Cultural Services								9.04	2.55	3.52	4.03
Health Services								0.02	0.02	0.03	0.03
All Services								10.12	10.77	10.00	9.44

Note: Import ratio is computed by dividing imports by gross productive output in the corresponding service sector. Imports and gross productive output data are secured from OECD databases.

Table 4.29 Import ratio: Sweden

Industry	2000	2001	2002	2003	2004	2005	2006	2007	2008	2009	2010
Telecommunication Services	8.59	7.67	5.52	7.75	10.85	11.52	12.34	11.74	11.78	12.80	12.60
Construction	1.47	1.31	1.22	2.13	1.48	1.34	1.70	1.96	2.16	2.89	2.37
Air Transportation	35.14	40.92	36.54	46.80	36.64	43.00	49.74	56.57	59.40	55.78	62.23
Insurance	6.87	6.50	7.49	6.17	5.08	5.58	4.64	4.55	5.20	6.33	5.73
Financial Services	5.96	6.14	6.02	5.69	4.47	5.50	7.05	3.53	3.56	3.00	2.68
IT Services	10.56	8.84	9.05	9.94	9.31	9.33	13.67	15.47	15.45	15.81	13.51
Accounting & Legal Services	11.77	15.14	12.18	12.98	12.45	10.43	10.73	12.11	10.04	11.70	10.59
Advertising Services	17.00	20.34	16.00	15.15	14.57	14.02	13.04	17.87	19.12	19.51	18.43
Education							0.02	0.04	0.05	0.06	0.05
Personal & Cultural Services	1.83	1.93	1.37	1.75	2.09	2.01	1.70	2.58	2.91	3.63	3.19
Health Services							0.02	0.04	0.05	0.05	0.02
All Services	8.76	9.26	8.55	8.22	8.26	8.50	8.75	9.08	9.54	9.44	8.82

Note: Import ratio is computed by dividing imports by gross productive output in the corresponding service sector. Imports and gross productive output data are secured from OECD databases.

Table 4.30 Import ratio: United States

Industry	2000	2001	2002	2003	2004	2005	2006	2007	2008	2009	2010
Telecommunication Services	1.07	0.92	0.81	0.81	0.85	0.79	1.07	1.18	1.22	1.21	1.25
Construction	0.14	0.18	0.25	0.18	0.19	0.09	0.12	0.19	0.27	0.33	0.24
Air Transportation	22.52	23.67	23.70	21.33	24.26	24.04	23.47	22.74	23.63	22.26	22.72
Insurance	2.61	3.71	4.66	4.92	5.19	4.78	6.32	7.19	8.88	10.23	9.78
Financial Services	1.95	1.66	1.33	1.23	1.47	1.45	1.66	2.06	1.76	1.42	1.43
IT Services	2.34	2.42	2.46	2.87	3.12	3.57	4.16	4.19	4.37	4.77	5.12
Education	0.31	0.33	0.37	0.41	0.44	0.46	0.49	0.49	0.50	0.50	0.52
Personal & Cultural Services	0.06	0.11	0.07	0.07	0.07	0.09	0.09	0.12	0.13	0.17	0.25
All Services	1.78	1.71	1.73	1.76	1.91	1.90	1.99	2.06	2.20	2.14	2.16

Note: Import ratio is computed by dividing imports by gross productive output in the corresponding service sector. Imports and gross productive output data are secured from OECD databases.

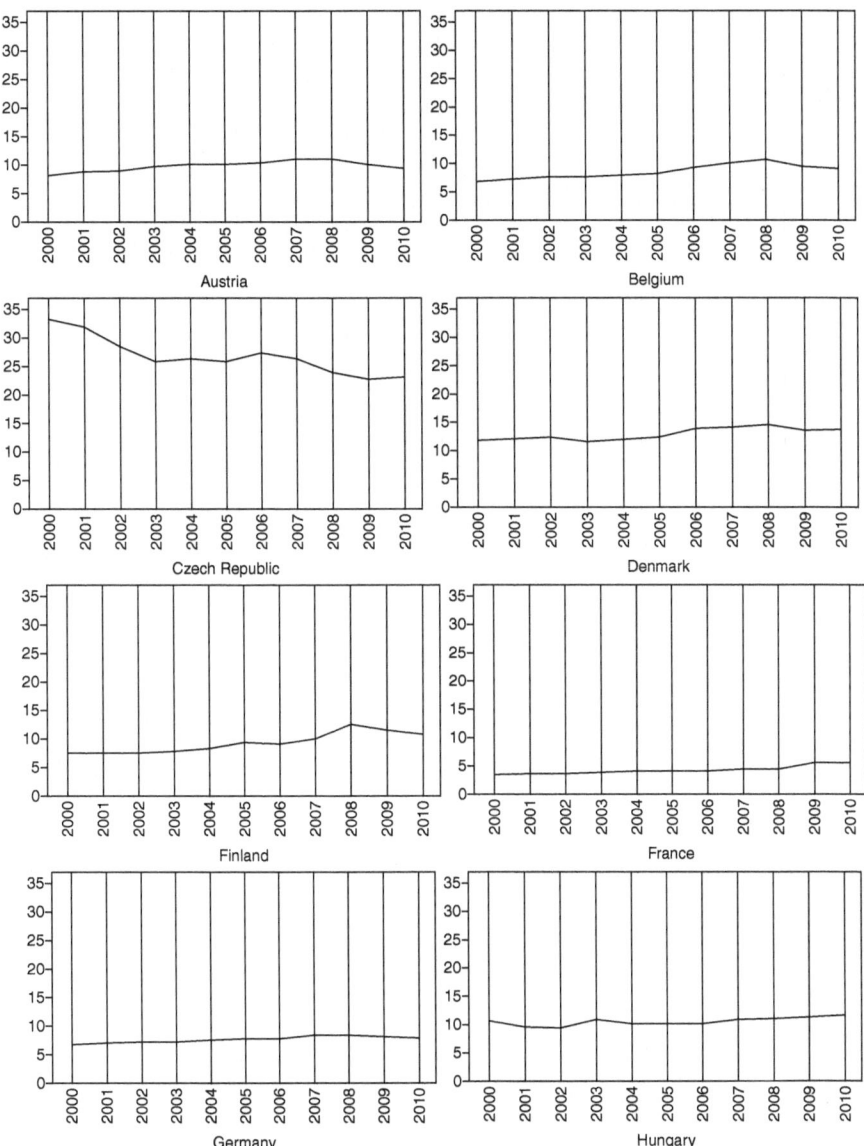

Fig. 4.2 Import ratio for service sector across countries (2000–2010)

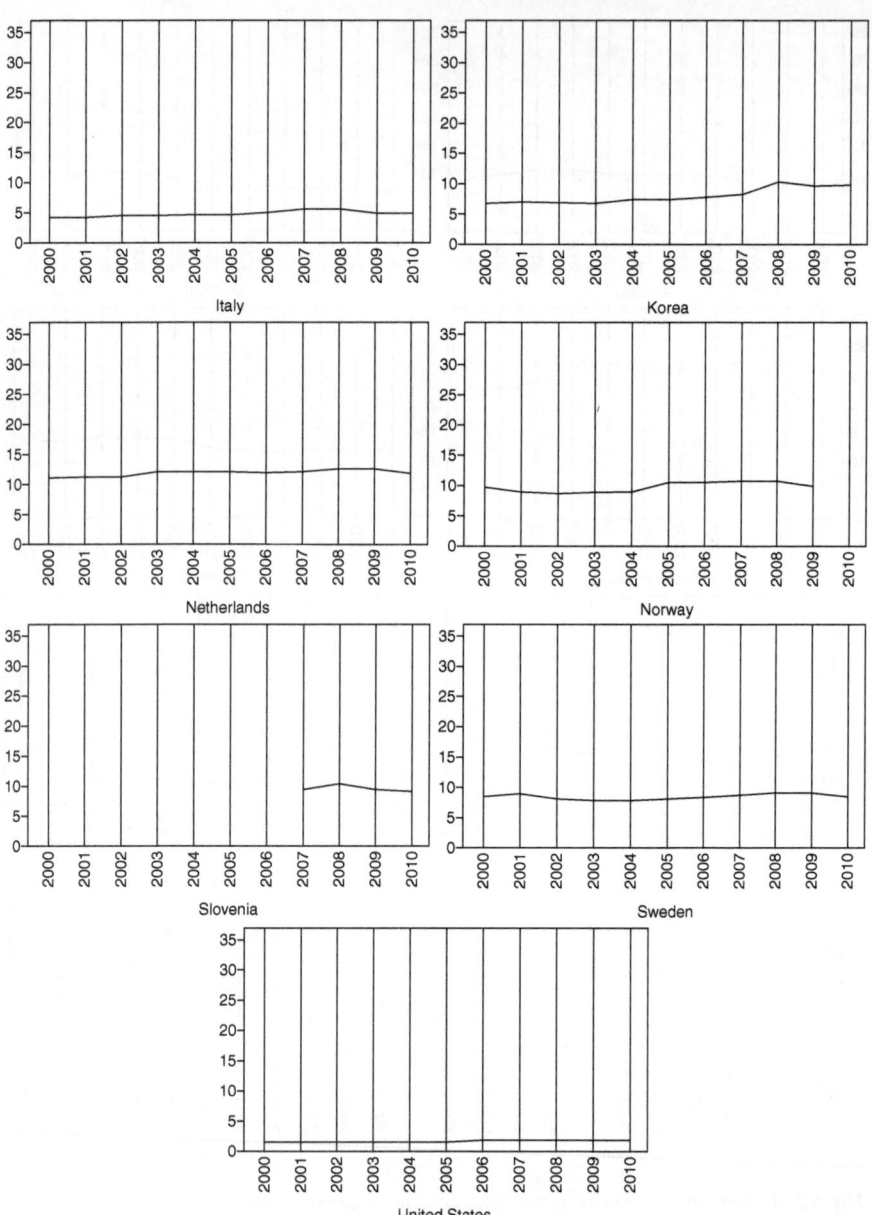

Fig. 4.2 (continued)

A overview of Import Ratio across countries during the years 2000–2010 across the fifteen OECD countries indicates that the Czech Republic's service economy is the most import-oriented (27.23), followed by Denmark (13.42), across service sectors. On average, the Czech Republic's Import Ratio is about twice that of Demark's, although it should be noted that the Czech Republic's imports has shown a declining trend during this decade. Part of the decline in numbers can be explained by the fact that the Czech Republic has undergone rapid economic changes as an newly independent country. The United States (1.94 %) is the least import-oriented among the nations for which data is provided in this book, followed by France (7.89).

As noted earlier with Export Ratio, much caution is needed while interpreting these numbers, as the context is critical. Import Ratio captures the extent to which an industry's products are imported in a country, but it does not capture the actual volume of exports from a country. The failure to recognize this difference could create a misleading picture if all facts are not carefully taken into consideration. For instance, as noted earlier, the Czech Republic imports a greater portion of the services in an industry compared to the United States. However, it should be noted that the United States is one of the major importers for services in aggregate value terms globally, which in many ways would dwarf the imports of the Czech Republic. The obvious reason for the United States having a lower Import Ratio number is the fact that the U.S. economy is the largest in many, if not most, of the service industry categories in the world [see the discussion on Domestic Global Ratio (Section 4.4) in this regard]. Therefore, the reader is advised to review the aggregate numbers to understand the context of the ratios presented. While in the case of the United States, the aggregate numbers for imports are presented in Chapter 2 of the book, for other countries, the reader should refer to another resource, such as the OECD databases.

4.3   Gross Profit Ratio

Gross Profit Ratio refers to the ratio of gross margins made by a firm on its sales. Gross profit figures are a useful source of information, indicating the surplus left for other expenses and profits after cost of production of the service has been paid for. Gross Profit Ratio is measured by dividing gross profit by sales revenue, while gross profit is measured as the sales revenue minus the cost of goods sold. Gross Profit Ratio is an important number, as it indicates the extent to which a firm can make profits and thus serves as a means of deciphering the returns possible in an industry. Gross Profit Ratio was computed based on numerical figures accessed from OECD databases as follows: Initially, gross surplus from a particular service sector and gross productive output were determined from the OECD database for each service industry segment in a country. Then, gross surplus was divided by gross productive output in the corresponding service sector and multiplied by 100 to arrive at the Gross Profit Ratio.

Gross Profit Ratio numbers for the years 2000–2010 for 15 OECD countries across 11–12 industries are presented where data was available. Table 4.31 shows the Gross Profit Ratio for Austria; Table 4.32 shows the Gross Profit Ratio for Belgium; Table 4.33 shows the Gross Profit Ratio for the Czech Republic; Table 4.34 shows the Gross Profit Ratio for Denmark; Table 4.35 shows the Gross Profit Ratio for Finland; Table 4.36 shows the Gross Profit Ratio for France; Table 4.37 shows the Gross Profit Ratio for Germany; Table 4.38 shows the Gross Profit Ratio for Hungary; Table 4.39 shows the Gross Profit Ratio for Italy; Table 4.40 shows the Gross Profit Ratio for Korea; Table 4.41 shows the Gross Profit Ratio for the Netherlands; Table 4.42 shows the Gross Profit Ratio for Norway; Table 4.43 shows the Gross Profit Ratio for Slovenia; Table 4.44 shows the Gross Profit Ratio for Sweden; and Table 4.45 shows the Gross Profit Ratio for the United States. Figure 4.3 graphically shows the average Gross Profit Ratio for the services sector in each of the countries.

Italy had the highest average (31.80) Gross Profit Ratio across service sectors in the years 2000–2010 for the 15 OECD countries, followed by Germany (27.71). Belgium had the lowest Gross Profit Ratio (16.88) in the years 2000–2010 for the 15 OECD countries, followed by Germany (17.60). The large variation indicates there exists a significant variation of gross profit across countries. Chapter 6 of this book presents underlying drivers which are causes for this variation.

Table 4.31 Gross profit ratio: Austria

Industry	2000	2001	2002	2003	2004	2005	2006	2007	2008	2009	2010
Construction (ISIC Rev.4 Code: 41)	17.30	16.96	17.42	19.29	19.66	19.40	17.95	18.59	17.51	15.74	15.56
Automotive Retail & Repair (ISIC Rev.4 Code: 45)	20.34	15.77	14.26	14.55	18.05	16.79	14.35	16.14	17.51	14.51	17.47
Wholesale Trade (ISIC Rev.4 Code: 46)	24.86	25.06	25.43	25.48	26.72	27.96	30.80	30.11	29.31	27.51	26.52
Retail Trade (ISIC Rev.4 Code: 47)	23.12	22.72	22.69	21.43	21.84	22.97	21.44	22.46	22.17	23.08	26.14
Land Transport (ISIC Rev.4 Code: 49)	25.20	25.80	26.67	24.63	21.08	17.70	16.76	17.19	16.08	16.50	14.55
Water Transport (ISIC Rev.4 Code: 50)	14.73	24.71	20.96	22.28	25.18	15.03	1.65	2.80	4.47	6.31	10.70
Air Transport (ISIC Rev.4 Code: 51)	8.28	3.50	9.28	11.17	11.80	5.23	8.60	10.93	4.96	5.44	9.97
Support Activities for Transport (ISIC Rev.4 Code: 52)	32.73	32.53	31.85	28.44	34.44	13.26	13.23	15.17	17.03	15.37	15.74
Postal & Courier Activities (ISIC Rev.4 Code: 53)	−1.40	−1.26	−1.43	3.33	5.94	11.82	11.72	10.55	9.38	8.70	8.44
Hotel & Restaurants (ISIC Rev.4 Code: 55, 56)	36.68	37.56	37.26	38.49	38.74	39.16	39.99	39.85	40.77	40.72	40.52
Publishing Activities (ISIC Rev.4 Code: 58)	11.61	15.44	16.86	14.85	15.73	16.32	15.31	14.88	14.51	12.32	12.82
Audiovisual & Broadcasting (ISIC Rev.4 Code: 59, 60)	18.92	17.34	15.53	16.69	19.54	19.98	16.59	16.41	14.43	14.24	15.61
Telecommunications (ISIC Rev.4 Code: 61)	15.95	24.22	28.84	31.49	25.97	26.81	27.10	26.50	26.33	26.51	28.34

(continued)

Table 4.31 (continued)

Industry	2000	2001	2002	2003	2004	2005	2006	2007	2008	2009	2010
IT & Information Services (ISIC Rev.4 Code: 62, 63)	16.38	16.88	17.32	15.55	14.09	16.63	15.85	14.94	15.58	13.75	13.44
Financial Service Activities (ISIC Rev.4 Code: 64)	24.92	23.23	23.99	21.24	22.57	19.41	24.35	25.52	25.75	18.11	25.88
Insurance & Pensions (ISIC Rev.4 Code: 65)	18.40	13.87	18.89	15.58	18.02	23.19	18.44	21.22	20.72	18.13	9.94
Auxiliary Financial Services (ISIC Rev.4 Code: 66)	20.75	17.87	17.97	18.98	16.26	18.05	14.72	13.15	6.05	8.39	9.60
Legal, Accounting & Consultancy (ISIC Rev.4 Code: 69,70)	28.60	28.93	28.66	26.43	26.00	26.01	24.36	20.83	22.11	22.74	22.15
Architectural & Technical Services (ISIC Rev.4 Code: 71)	13.83	15.11	17.17	16.37	16.32	17.10	18.19	19.57	19.93	17.43	18.37
Research & Development (ISIC Rev.4 Code: 72)	−19.14	−15.83	−8.81	−1.71	2.23	0.55	−14.82	−10.76	−0.19	−3.65	−6.56
Advertising & Market Research (ISIC Rev.4 Code: 73)	7.27	8.57	9.92	7.74	8.12	8.00	9.78	10.05	11.16	9.96	9.50
Other Professionals Services (ISIC Rev.4 Code: 74, 75)	27.72	22.41	24.98	22.04	21.37	21.75	23.16	22.01	26.33	27.66	28.35
Rental & Leasing Activities (ISIC Rev.4 Code: 77)	61.65	61.03	60.70	62.05	62.21	61.87	62.51	62.59	61.13	60.87	61.01
Employment Activities (ISIC Rev.4 Code: 78)	2.79	3.53	5.01	3.73	5.73	4.98	6.70	9.14	5.33	3.80	5.10
Travel Service (ISIC Rev.4 Code: 79)	4.99	6.24	5.15	5.09	4.58	1.83	−1.85	4.75	1.87	5.12	4.68
Security & Building Services (ISIC Rev.4 Code: 80, 81, 82)	27.86	26.17	25.91	24.02	23.07	23.55	21.30	22.34	22.53	23.15	24.88

Public Administration & Defense (ISIC Rev.4 Code: 84)	10.23	10.07	10.10	9.71	9.81	9.63	9.43	9.44	9.11	9.27	9.51
Education (ISIC Rev.4 Code: 85)	7.08	7.66	7.36	7.24	7.25	6.70	7.10	7.04	6.65	6.19	6.03
Human Health Activities (ISIC Rev.4 Code: 86)	7.85	10.85	11.67	14.99	15.79	16.32	18.03	17.05	17.92	15.82	15.39
Residential Care & Social Work (ISIC Rev.4 Code: 87, 88)	0.68	−5.15	−5.62	−5.69	−6.12	−6.65	−6.97	−8.09	−8.14	−4.86	−6.30
Creative Entertainment Activities (ISIC Rev.4 Code: 90, 91, 92)	24.16	22.84	25.22	25.97	26.09	29.90	27.11	26.97	29.83	28.40	26.30
Sports & Recreation Activities (ISIC Rev.4 Code: 93)	34.79	34.85	35.90	35.70	35.08	34.07	34.12	36.46	34.72	33.00	34.28
Membership Organizations (ISIC Rev.4 Code: 94)	6.04	6.24	6.48	6.37	6.46	6.45	6.78	6.95	5.94	5.57	5.75
Repair Services (ISIC Rev.4 Code: 95)	20.59	16.91	17.30	11.29	11.13	7.58	17.37	25.57	25.89	26.83	20.80
Personal Service Activities (ISIC Rev.4 Code: 96)	38.56	38.90	39.86	40.04	40.67	41.57	41.04	41.24	40.84	41.64	42.47
Undifferentiated Services (ISIC Rev.4 Code: 97, 98)											
All Services (Average)	24.27	24.28	24.93	24.82	25.33	25.15	25.47	25.47	25.26	24.58	24.92
Business Sector Services (Average)	29.21	29.06	29.76	29.44	29.93	29.53	29.79	29.80	29.62	29.16	29.64
Business Sector Services (Average excluding real estate)	23.56	23.50	24.25	23.74	24.00	23.38	23.70	23.96	23.87	22.85	23.41
All Industries (Average)	21.59	21.44	21.81	21.73	22.12	22.02	21.98	21.89	20.85	20.43	20.62

Note: Gross Profit Ratio is computed by dividing gross surplus by gross productive output in the corresponding service sector. Gross surplus and gross productive output data are secured from OECD databases.

Table 4.32 Gross profit ratio: Belgium

Industry	2000	2001	2002	2003	2004	2005	2006	2007	2008	2009	2010
Construction (ISIC Rev.4 Code: 41)	13.64	13.53	13.55	13.71	14.02	14.05	14.20	14.82	14.82	14.14	13.79
Automotive Retail & Repair (ISIC Rev.4 Code: 45)	17.90	17.93	15.47	17.85	19.12	19.08	18.97	20.32	15.75	15.20	18.26
Wholesale Trade (ISIC Rev.4 Code: 46)	15.61	15.25	16.46	18.59	20.09	19.51	18.62	19.27	19.01	17.87	17.81
Retail Trade (ISIC Rev.4 Code: 47)	17.75	17.66	18.30	21.98	23.07	23.70	23.91	24.43	23.25	22.74	21.76
Land Transport (ISIC Rev.4 Code: 49)	12.61	13.00	11.09	11.47	9.23	9.84	10.54	11.04	9.12	8.39	8.81
Water Transport (ISIC Rev.4 Code: 50)	4.07	2.11	-1.27	0.58	7.07	17.39	14.27	13.19	15.97	9.36	13.61
Air Transport (ISIC Rev.4 Code: 51)	-3.72	-4.73	-2.47	3.73	5.80	6.04	6.91	6.25	5.00	4.91	0.54
Support Activities for Transport (ISIC Rev.4 Code: 52)	19.71	19.06	19.00	19.14	19.25	20.48	20.57	20.68	20.00	20.99	19.67
Postal & Courier Activities (ISIC Rev.4 Code: 53)	10.51	5.10	1.49	8.06	9.04	7.05	9.62	11.15	11.53	12.47	13.99
Hotel & Restaurants (ISIC Rev.4 Code: 55, 56)	18.24	17.63	16.95	16.85	16.81	16.68	16.14	16.15	15.88	14.47	15.48
Publishing Activities (ISIC Rev.4 Code: 58)	14.41	11.24	11.88	13.74	14.67	17.22	16.40	17.70	17.36	20.77	19.55
Audiovisual & Broadcasting (ISIC Rev.4 Code: 59, 60)	14.79	15.12	14.96	14.57	14.12	13.87	16.29	18.04	18.60	18.47	20.95
Telecommunications (ISIC Rev.4 Code: 61)	26.58	24.03	27.63	28.37	32.11	33.42	32.61	33.15	32.15	32.66	34.46

Industry											
IT & Information Services (ISIC Rev.4 Code: 62, 63)	9.98	13.84	13.57	14.72	15.57	17.33	16.93	18.07	17.82	16.46	17.74
Financial Service Activities (ISIC Rev.4 Code: 64)	19.63	16.21	19.67	16.72	19.64	20.05	18.85	17.01	14.01	22.58	28.03
Insurance & Pensions (ISIC Rev.4 Code: 65)	13.79	13.39	16.58	19.76	21.54	16.13	16.56	14.41	17.04	19.42	19.92
Auxiliary Financial Services (ISIC Rev.4 Code: 66)	20.75	20.75	22.13	23.07	24.99	25.18	23.37	22.96	17.60	18.33	20.34
Legal, Accounting & Consultancy (ISIC Rev.4 Code: 69,70)	43.57	39.95	39.33	41.06	39.32	38.57	38.37	37.38	37.93	39.48	38.95
Architectural & Technical Services (ISIC Rev.4 Code: 71)	16.71	16.04	15.52	19.35	19.79	19.43	20.31	22.19	21.03	20.29	20.10
Research & Development (ISIC Rev.4 Code: 72)	10.57	9.61	10.53	13.34	10.21	9.98	8.99	7.62	8.47	10.10	4.48
Advertising & Market Research (ISIC Rev.4 Code: 73)	9.39	9.07	8.28	10.21	10.84	10.65	11.36	12.91	11.51	9.24	10.37
Other Professionals Services (ISIC Rev.4 Code: 74, 75)	31.88	31.56	29.51	31.24	31.50	29.60	30.91	28.80	29.15	28.54	27.00
Rental & Leasing Activities (ISIC Rev.4 Code: 77)	40.09	37.26	39.04	39.00	40.12	38.56	38.71	40.39	40.94	40.79	40.54
Employment Activities (ISIC Rev.4 Code: 78)	4.27	3.43	2.78	2.96	3.39	4.07	2.68	2.20	1.06	1.39	3.23
Travel Service (ISIC Rev.4 Code: 79)	2.17	4.11	7.49	5.61	5.40	5.33	4.52	5.76	5.39	5.02	5.46
Security & Building Services (ISIC Rev.4 Code: 80, 81, 82)	18.30	16.59	17.89	16.74	18.85	18.58	19.40	18.94	18.09	17.49	18.42
Public Administration & Defense (ISIC Rev.4 Code: 84)	6.16	6.13	5.75	5.70	5.68	5.89	6.17	6.42	6.48	6.28	6.49

(continued)

Table 4.32 (continued)

Industry	2000	2001	2002	2003	2004	2005	2006	2007	2008	2009	2010
Education (ISIC Rev.4 Code: 85)	6.38	6.14	6.03	5.88	5.92	5.85	5.98	5.95	5.76	5.24	5.32
Human Health Activities (ISIC Rev.4 Code: 86)	24.05	22.92	21.68	22.67	22.79	22.77	22.50	22.66	23.13	23.21	22.65
Residential Care & Social Work (ISIC Rev.4 Code: 87, 88)	10.55	10.17	10.12	6.69	8.21	7.93	8.63	8.54	7.46	6.36	5.72
Creative Entertainment Activities (ISIC Rev.4 Code: 90, 91, 92)	21.32	19.09	17.94	20.00	21.60	21.33	20.87	20.69	18.20	16.44	17.05
Sports & Recreation Activities (ISIC Rev.4 Code: 93)	22.63	17.86	17.92	20.67	21.72	22.10	19.26	20.92	19.68	21.04	19.47
Membership Organizations (ISIC Rev.4 Code: 94)	11.42	8.76	8.97	7.13	7.27	7.12	6.17	6.19	5.59	3.90	4.53
Repair Services (ISIC Rev.4 Code: 95)	29.98	30.62	31.22	20.65	19.90	21.06	17.93	21.80	22.01	20.97	20.45
Personal Service Activities (ISIC Rev.4 Code: 96)	28.41	28.88	27.81	28.71	28.63	27.82	29.57	28.84	26.83	26.63	25.32
Undifferentiated Services (ISIC Rev.4 Code: 97, 98)											
All Services (Average)											
Business Sector Services (Average)											
Business Sector Services (Average excluding real estate)											
All Industries (Average)	17.50	16.91	17.27	17.84	18.18	18.35	18.02	18.21	17.51	18.16	18.26

Note: Gross Profit Ratio is computed by dividing gross surplus by gross productive output in the corresponding service sector. Gross surplus and gross productive output data are secured from OECD databases.

Table 4.33 Gross profit ratio: Czech Republic

Industry	2000	2001	2002	2003	2004	2005	2006	2007	2008	2009	2010
Construction (ISIC Rev.4 Code: 41)	14.33	15.25	15.33	15.47	16.66	16.12	14.43	14.21	14.09	16.16	17.23
Automotive Retail & Repair (ISIC Rev.4 Code: 45)	28.02	27.88	27.17	25.91	23.23	22.05	23.31	23.23	20.33	14.10	21.25
Wholesale Trade (ISIC Rev.4 Code: 46)	29.97	30.64	30.92	29.86	27.19	29.68	29.88	28.31	26.21	22.02	22.00
Retail Trade (ISIC Rev.4 Code: 47)	32.10	35.71	34.45	31.40	29.00	30.92	31.35	30.63	27.99	27.79	25.23
Land Transport (ISIC Rev.4 Code: 49)	30.87	30.04	33.29	26.76	24.85	22.35	24.14	21.97	19.77	18.88	18.66
Water Transport (ISIC Rev.4 Code: 50)	23.11	19.47	20.31	18.00	9.74	9.71	8.90	9.71	6.80	6.88	4.11
Air Transport (ISIC Rev.4 Code: 51)	20.25	19.56	22.39	25.83	19.36	10.82	10.25	11.05	11.36	7.04	9.41
Support Activities for Transport (ISIC Rev.4 Code: 52)	30.07	28.11	24.91	27.55	26.34	27.77	25.66	24.70	24.18	26.74	27.26
Postal & Courier Activities (ISIC Rev.4 Code: 53)	13.00	9.03	7.10	7.17	5.72	3.41	3.38	6.49	7.38	6.87	6.36
Hotel & Restaurants (ISIC Rev.4 Code: 55, 56)	36.92	34.77	33.52	33.48	32.57	28.07	24.65	23.70	22.67	20.77	22.04
Publishing Activities (ISIC Rev.4 Code: 58)	12.82	12.20	12.16	11.87	11.03	13.99	13.42	13.23	12.79	11.83	15.25
Audiovisual & Broadcasting (ISIC Rev.4 Code: 59, 60)	24.75	23.52	27.35	37.45	34.76	35.42	33.87	34.51	35.24	30.40	31.77
Telecommunications (ISIC Rev.4 Code: 61)	39.20	39.82	41.47	40.09	39.63	39.68	41.15	42.13	41.87	41.40	41.33

(continued)

Table 4.33 (continued)

Industry	2000	2001	2002	2003	2004	2005	2006	2007	2008	2009	2010
IT & Information Services (ISIC Rev.4 Code: 62, 63)	26.23	21.29	24.07	23.06	19.08	22.98	21.78	21.71	20.77	20.81	19.96
Financial Service Activities (ISIC Rev.4 Code: 64)	30.75	29.73	15.23	26.14	26.49	21.03	18.55	23.96	29.76	38.81	40.64
Insurance & Pensions (ISIC Rev.4 Code: 65)	−15.76	10.60	15.72	11.97	7.38	−0.73	18.39	20.67	28.88	22.38	22.04
Auxiliary Financial Services (ISIC Rev.4 Code: 66)	17.78	24.81	26.88	31.75	46.82	40.70	27.10	28.81	32.40	35.54	36.17
Legal, Accounting & Consultancy (ISIC Rev.4 Code: 69,70)	32.41	29.80	34.49	32.03	37.60	40.43	38.36	32.73	30.43	26.88	26.56
Architectural & Technical Services (ISIC Rev.4 Code: 71)	17.45	19.88	21.58	18.98	17.90	21.02	20.13	19.54	20.16	19.53	19.37
Research & Development (ISIC Rev.4 Code: 72)	23.15	30.48	17.64	18.10	18.48	18.64	17.83	19.21	19.61	18.77	18.58
Advertising & Market Research (ISIC Rev.4 Code: 73)	11.60	10.50	15.28	13.82	15.15	15.43	12.87	8.47	9.14	8.81	9.24
Other Professionals Services (ISIC Rev.4 Code: 74, 75)	17.82	27.41	29.82	21.31	21.42	22.31	23.30	20.29	20.74	23.35	28.71
Rental & Leasing Activities (ISIC Rev.4 Code: 77)	41.23	36.44	37.28	40.28	37.01	37.59	37.27	39.15	39.74	42.85	42.56
Employment Activities (ISIC Rev.4 Code: 78)	12.88	9.70	20.29	23.15	17.30	19.41	14.67	10.87	18.09	21.59	8.96
Travel Service (ISIC Rev.4 Code: 79)	8.73	10.53	9.18	8.17	7.66	6.48	7.35	6.65	6.39	4.51	6.11
Security & Building Services (ISIC Rev.4 Code: 80, 81, 82)	21.43	23.53	22.30	19.31	19.56	16.51	19.55	23.72	23.86	16.41	18.04

Public Administration & Defense (ISIC Rev.4 Code: 84)	22.43	21.48	22.98	21.52	22.03	21.69	21.73	21.98	22.01	22.51	23.17
Education (ISIC Rev.4 Code: 85)	23.46	22.54	21.25	19.15	19.35	19.33	18.65	18.73	19.31	18.24	18.72
Human Health Activities (ISIC Rev.4 Code: 86)	21.33	23.22	23.99	22.37	21.92	22.56	19.99	18.30	18.31	20.29	17.51
Residential Care & Social Work (ISIC Rev.4 Code: 87, 88)	28.08	25.29	22.62	19.47	21.15	19.54	18.44	18.30	17.18	17.29	12.19
Creative Entertainment Activities (ISIC Rev.4 Code: 90, 91, 92)	27.16	27.21	27.01	25.69	25.43	25.04	23.92	22.10	23.58	25.62	25.25
Sports & Recreation Activities (ISIC Rev.4 Code: 93)	31.18	33.45	26.54	24.68	23.28	21.75	24.60	25.60	21.96	24.21	21.48
Membership Organizations (ISIC Rev.4 Code: 94)	4.10	1.80	1.64	6.30	-1.82	12.81	13.60	12.57	8.02	8.03	8.39
Repair Services (ISIC Rev.4 Code: 95)	30.86	33.77	34.69	33.27	30.33	28.57	28.61	24.96	29.06	30.64	28.22
Personal Service Activities (ISIC Rev.4 Code: 96)	69.41	67.08	69.09	66.18	66.00	60.37	63.16	63.89	56.66	57.24	62.12
Undifferentiated Services (ISIC Rev.4 Code: 97, 98)											
All Services (Average)	29.32	29.43	29.30	28.15	26.86	26.77	26.69	26.00	25.63	25.69	25.78
Business Sector Services (Average)	30.67	30.95	30.83	29.76	28.11	28.04	28.01	27.14	26.76	26.77	26.97
Business Sector Services (Average excluding real estate)	27.72	28.15	28.18	27.10	25.72	25.37	25.41	24.74	24.26	23.41	23.94
All Industries (Average)	21.27	21.22	21.08	20.64	20.22	20.16	19.65	19.42	19.47	20.98	20.19

Note: Gross Profit Ratio is computed by dividing gross surplus by gross productive output in the corresponding service sector. Gross surplus and gross productive output data are secured from OECD databases.

Table 4.34 Gross profit ratio: Denmark

Industry	2000	2001	2002	2003	2004	2005	2006	2007	2008	2009	2010
Construction (ISIC Rev.4 Code: 41)	10.54	8.29	8.95	10.00	10.22	9.25	9.06	7.55	10.52	9.13	5.92
Automotive Retail & Repair (ISIC Rev.4 Code: 45)	12.78	14.07	13.30	12.43	15.43	13.57	10.82	10.69	12.13	8.76	8.72
Wholesale Trade (ISIC Rev.4 Code: 46)	16.46	14.66	15.88	15.93	15.87	12.72	15.94	14.76	15.38	12.78	17.06
Retail Trade (ISIC Rev.4 Code: 47)	19.07	16.97	18.02	17.74	17.06	16.78	17.17	16.55	8.50	7.54	9.05
Land Transport (ISIC Rev.4 Code: 49)	20.52	20.60	21.65	20.30	20.66	17.13	17.47	16.01	15.63	15.38	15.74
Water Transport (ISIC Rev.4 Code: 50)	13.31	11.77	11.84	12.84	17.87	19.08	8.09	7.30	6.54	-3.21	13.16
Air Transport (ISIC Rev.4 Code: 51)	1.23	1.79	-0.02	0.18	-5.13	-7.85	1.70	3.26	-12.37	1.53	8.57
Support Activities for Transport (ISIC Rev.4 Code: 52)	41.25	40.72	37.24	36.98	35.46	24.85	28.28	31.28	21.23	19.85	23.01
Postal & Courier Activities (ISIC Rev.4 Code: 53)	0.31	13.71	11.86	12.06	13.38	13.68	6.52	6.31	3.77	-0.96	1.61
Hotel & Restaurants (ISIC Rev.4 Code: 55, 56)	12.79	12.35	13.41	13.21	11.85	10.58	11.72	11.18	9.38	8.29	7.82
Publishing Activities (ISIC Rev.4 Code: 58)	11.55	10.52	9.61	11.40	9.34	10.38	7.78	4.60	7.70	6.85	11.11
Audiovisual & Broadcasting (ISIC Rev.4 Code: 59, 60)	18.74	13.78	19.07	18.80	13.81	13.37	8.88	5.46	5.17	1.85	5.72
Telecommunications (ISIC Rev.4 Code: 61)	28.07	19.08	18.93	24.08	22.79	23.22	23.36	24.77	25.13	25.24	21.86

IT & Information Services (ISIC Rev.4 Code: 62, 63)	5.24	5.24	2.00	4.02	3.37	-0.01	0.61	0.83	0.94	5.02	6.69
Financial Service Activities (ISIC Rev.4 Code: 64)	22.79	21.35	24.64	26.14	25.80	24.65	20.53	19.20	22.83	24.14	27.07
Insurance & Pensions (ISIC Rev.4 Code: 65)	14.96	6.99	7.20	16.04	20.64	20.13	24.08	20.46	21.82	18.93	6.15
Auxiliary Financial Services (ISIC Rev.4 Code: 66)	25.70	27.06	20.97	14.29	15.88	17.68	24.21	25.21	28.62	28.95	26.45
Legal, Accounting & Consultancy (ISIC Rev.4 Code: 69,70)	28.08	24.57	19.39	18.57	23.96	23.13	18.45	15.19	15.71	14.82	18.90
Architectural & Technical Services (ISIC Rev.4 Code: 71)	15.65	16.14	11.90	7.45	9.10	8.50	10.94	9.82	7.10	6.10	8.22
Research & Development (ISIC Rev.4 Code: 72)	9.06	7.55	15.49	5.44	-0.26	-2.81	-2.43	-9.01	22.39	27.64	28.30
Advertising & Market Research (ISIC Rev.4 Code: 73)	8.62	8.28	6.10	5.77	8.55	7.27	7.46	5.44	8.00	5.37	5.10
Other Professionals Services (ISIC Rev.4 Code: 74, 75)	30.14	26.91	25.03	25.73	28.69	34.05	28.45	17.16	18.75	18.50	18.76
Rental & Leasing Activities (ISIC Rev.4 Code: 77)	29.89	25.12	27.48	29.01	27.28	22.34	19.54	17.04	19.72	21.69	21.96
Employment Activities (ISIC Rev.4 Code: 78)	2.49	3.47	2.23	1.72	1.57	5.88	3.30	2.74	8.26	-0.71	-1.71
Travel Service (ISIC Rev.4 Code: 79)	-14.40	-20.00	-20.13	-15.22	-11.92	-4.23	-10.42	-0.05	-1.06	8.08	7.37
Security & Building Services (ISIC Rev.4 Code: 80, 81, 82)	21.13	19.32	18.29	17.75	17.95	20.91	19.78	16.61	9.29	11.91	13.35
Public Administration & Defense (ISIC Rev.4 Code: 84)	10.38	9.93	10.00	9.66	9.94	9.87	10.20	10.48	10.79	10.07	10.18

(continued)

Table 4.34 (continued)

Industry	2000	2001	2002	2003	2004	2005	2006	2007	2008	2009	2010
Education (ISIC Rev.4 Code: 85)	8.28	7.86	7.90	7.62	7.40	6.92	6.89	6.98	6.98	6.23	5.67
Human Health Activities (ISIC Rev.4 Code: 86)	13.70	13.19	12.93	12.97	12.97	12.59	12.41	12.28	13.07	13.74	13.40
Residential Care & Social Work (ISIC Rev.4 Code: 87, 88)	3.32	3.05	2.89	2.86	2.70	2.71	2.51	2.62	1.56	1.25	0.82
Creative Entertainment Activities (ISIC Rev.4 Code: 90, 91, 92)	44.34	45.47	41.14	37.85	39.63	42.69	40.16	37.86	41.11	38.89	39.21
Sports & Recreation Activities (ISIC Rev.4 Code: 93)	15.50	14.77	19.14	17.20	15.42	3.36	9.51	8.81	9.19	8.52	10.75
Membership Organizations (ISIC Rev.4 Code: 94)	8.33	6.96	6.39	6.31	6.04	7.15	1.51	1.63	0.48	-0.40	0.94
Repair Services (ISIC Rev.4 Code: 95)	5.30	4.09	5.19	5.62	6.55	9.69	14.42	11.50	5.12	24.57	29.51
Personal Service Activities (ISIC Rev.4 Code: 96)	38.72	37.97	35.69	36.25	36.94	35.01	35.04	30.96	32.66	34.08	35.92
Undifferentiated Services (ISIC Rev.4 Code: 97, 98)											
All Services (Average)	20.60	19.56	19.36	19.54	19.77	18.81	17.75	16.86	16.34	16.07	17.96
Business Sector Services (Average)	24.56	23.23	23.11	23.55	23.86	22.35	20.81	19.54	18.83	18.92	21.57
Business Sector Services (Average excluding real estate)	17.03	15.63	15.62	16.20	16.85	15.54	14.31	13.01	12.49	11.36	14.67
All Industries (Average)	19.84	18.74	18.56	18.55	18.91	18.34	17.73	16.62	16.15	15.47	17.43

Note: Gross Profit Ratio is computed by dividing gross surplus by gross productive output in the corresponding service sector. Gross surplus and gross productive output data are secured from OECD databases.

Table 4.35 Gross profit ratio: Finland

Industry	2000	2001	2002	2003	2004	2005	2006	2007	2008	2009	2010
Construction (ISIC Rev.4 Code: 41)	15.14	14.51	13.05	14.26	15.29	16.24	16.21	15.42	14.08	13.58	11.91
Automotive Retail & Repair (ISIC Rev.4 Code: 45)	22.98	25.38	24.73	29.16	27.70	27.18	20.31	21.67	25.39	21.13	24.17
Wholesale Trade (ISIC Rev.4 Code: 46)	17.72	18.64	19.37	17.76	18.00	17.04	16.70	17.90	20.83	17.39	17.67
Retail Trade (ISIC Rev.4 Code: 47)	18.94	22.52	21.87	19.69	22.38	21.18	19.71	20.71	19.34	17.30	17.19
Land Transport (ISIC Rev.4 Code: 49)	28.90	32.16	30.42	27.75	24.90	22.85	21.53	21.14	20.64	19.37	20.05
Water Transport (ISIC Rev.4 Code: 50)	20.93	20.50	22.53	24.56	25.61	23.72	23.43	23.80	13.91	6.34	8.99
Air Transport (ISIC Rev.4 Code: 51)	23.12	21.57	24.25	20.79	24.08	23.30	18.83	20.10	9.22	7.84	9.77
Support Activities for Transport (ISIC Rev.4 Code: 52)	19.30	21.73	19.32	17.32	16.91	15.19	8.76	8.86	11.39	11.05	10.50
Postal & Courier Activities (ISIC Rev.4 Code: 53)	17.18	17.03	18.08	18.00	16.01	16.72	15.57	14.03	9.21	10.90	9.55
Hotel & Restaurants (ISIC Rev.4 Code: 55, 56)	5.96	9.87	9.81	9.78	12.16	12.32	13.54	15.18	13.07	10.56	11.26
Publishing Activities (ISIC Rev.4 Code: 58)	12.35	13.90	14.71	13.82	16.66	14.75	14.21	17.41	19.53	19.34	17.66
Audiovisual & Broadcasting (ISIC Rev.4 Code: 59, 60)	25.86	28.10	25.30	16.94	25.46	22.44	19.21	21.15	21.08	20.21	21.74
Telecommunications (ISIC Rev.4 Code: 61)	27.64	30.00	33.24	29.97	29.21	20.27	20.07	28.17	27.91	29.81	30.11

(continued)

Table 4.35 (continued)

Industry	2000	2001	2002	2003	2004	2005	2006	2007	2008	2009	2010
IT & Information Services (ISIC Rev.4 Code: 62, 63)	12.07	13.59	10.52	11.90	16.65	16.11	13.74	17.85	14.24	11.65	12.57
Financial Service Activities (ISIC Rev.4 Code: 64)	35.96	37.95	31.89	27.44	27.16	29.72	32.68	33.41	27.01	24.85	19.24
Insurance & Pensions (ISIC Rev.4 Code: 65)	64.80	48.30	54.68	10.32	25.51	9.79	22.33	38.37	33.61	19.41	30.50
Auxiliary Financial Services (ISIC Rev.4 Code: 66)	26.28	24.19	21.67	16.12	22.86	21.04	24.05	26.40	17.95	13.14	15.36
Legal, Accounting & Consultancy (ISIC Rev.4 Code: 69,70)	18.37	19.64	17.93	19.09	22.72	22.14	19.47	22.13	20.17	20.11	19.44
Architectural & Technical Services (ISIC Rev.4 Code: 71)	19.91	20.43	17.12	15.98	16.70	16.19	16.09	18.59	14.56	11.47	10.05
Research & Development (ISIC Rev.4 Code: 72)	7.03	2.66	0.68	3.94	-0.36	0.25	7.78	-6.27	4.66	3.79	5.02
Advertising & Market Research (ISIC Rev.4 Code: 73)	9.74	12.90	10.71	10.48	10.89	11.55	9.72	13.53	13.17	13.52	16.25
Other Professionals Services (ISIC Rev.4 Code: 74, 75)	16.67	23.51	21.80	20.11	16.78	16.63	20.35	20.88	20.90	19.67	18.68
Rental & Leasing Activities (ISIC Rev.4 Code: 77)	29.25	31.80	29.82	26.87	27.18	31.03	28.93	27.08	37.46	34.58	34.11
Employment Activities (ISIC Rev.4 Code: 78)	7.66	10.90	8.81	8.60	7.99	6.83	7.20	7.23	7.94	5.14	7.75
Travel Service (ISIC Rev.4 Code: 79)	10.87	9.35	10.20	11.16	11.81	13.20	12.96	13.56	13.87	7.25	6.67
Security & Building Services (ISIC Rev.4 Code: 80, 81, 82)	13.53	16.54	14.46	15.26	15.39	14.69	14.94	16.19	18.00	16.15	13.99

Public Administration & Defense (ISIC Rev.4 Code: 84)	14.60	14.36	13.77	13.27	13.36	13.59	13.57	13.51	13.29	12.56	12.67
Education (ISIC Rev.4 Code: 85)	8.08	7.89	7.42	7.18	7.11	7.23	7.08	7.82	7.42	7.17	6.96
Human Health Activities (ISIC Rev.4 Code: 86)	9.01	9.49	9.40	9.53	9.45	9.47	9.44	9.90	9.89	10.07	10.48
Residential Care & Social Work (ISIC Rev.4 Code: 87, 88)	5.62	5.99	5.89	5.98	6.05	6.12	5.57	5.77	5.99	6.01	6.45
Creative Entertainment Activities (ISIC Rev.4 Code: 90, 91, 92)	13.17	11.66	12.44	11.38	11.56	12.79	14.68	14.80	21.48	18.58	19.66
Sports & Recreation Activities (ISIC Rev.4 Code: 93)	25.48	27.61	26.88	24.83	23.36	22.20	21.01	20.52	17.94	17.99	16.50
Membership Organizations (ISIC Rev.4 Code: 94)	5.56	5.72	5.87	5.78	5.41	5.70	5.99	6.25	6.34	6.23	5.95
Repair Services (ISIC Rev.4 Code: 95)	22.79	23.10	19.08	2.65	9.16	18.27	22.71	16.27	18.72	19.46	19.69
Personal Service Activities (ISIC Rev.4 Code: 96)	32.65	34.39	34.05	36.01	37.21	36.96	40.58	38.25	36.78	40.09	40.85
Undifferentiated Services (ISIC Rev.4 Code: 97, 98)											
All Services (Average)	24.36	24.92	24.69	23.16	23.63	22.61	22.03	23.05	22.81	21.56	21.58
Business Sector Services (Average)	29.87	30.70	30.56	28.67	29.32	27.80	26.84	28.22	28.02	26.77	26.75
Business Sector Services (Average excluding real estate)	21.64	22.54	22.21	19.55	20.49	18.77	17.86	19.61	18.80	16.69	16.80
All Industries (Average)	21.28	21.81	21.90	21.12	21.12	20.16	19.79	20.70	19.63	18.54	18.78

Note: Gross Profit Ratio is computed by dividing gross surplus by gross productive output in the corresponding service sector. Gross surplus and gross productive output data are secured from OECD databases.

Table 4.36 Gross profit ratio: France

Industry	2000	2001	2002	2003	2004	2005	2006	2007	2008	2009	2010
Construction (ISIC Rev.4 Code: 41)	12.55	13.53	13.29	13.28	13.55	13.50	13.91	14.91	15.88	15.00	14.26
Automotive Retail & Repair (ISIC Rev.4 Code: 45)	17.19	17.51	17.19	15.89	15.92	15.79	15.35	17.35	17.98	16.10	17.35
Wholesale Trade (ISIC Rev.4 Code: 46)	19.65	21.27	18.48	19.85	16.18	15.33	15.09	14.87	16.94	15.35	16.88
Retail Trade (ISIC Rev.4 Code: 47)	22.93	24.66	25.38	24.05	24.06	22.26	21.27	21.56	21.79	21.81	20.15
Land Transport (ISIC Rev.4 Code: 49)	13.09	14.38	12.20	11.11	11.10	11.99	11.96	13.08	12.94	13.46	12.36
Water Transport (ISIC Rev.4 Code: 50)	1.65	2.27	1.32	1.20	6.86	8.03	4.96	5.63	7.28	−3.16	9.37
Air Transport (ISIC Rev.4 Code: 51)	−0.43	−2.77	0.27	−0.88	1.53	2.37	3.65	3.32	5.76	4.85	5.51
Support Activities for Transport (ISIC Rev.4 Code: 52)	21.18	19.91	20.06	19.68	20.75	20.31	20.70	21.39	22.17	23.56	21.37
Postal & Courier Activities (ISIC Rev.4 Code: 53)	−14.20	−12.88	−7.44	−4.86	−1.74	−2.43	−3.43	−1.51	−1.76	−2.91	0.15
Hotel & Restaurants (ISIC Rev.4 Code: 55, 56)	21.39	20.83	20.94	20.57	20.26	20.73	20.10	20.71	19.70	18.14	16.84
Publishing Activities (ISIC Rev.4 Code: 58)							13.99	12.78	12.70	11.07	12.40
Audiovisual & Broadcasting (ISIC Rev.4 Code: 59, 60)							21.19	18.60	17.17	17.68	17.62
Telecommunications (ISIC Rev.4 Code: 61)	23.26	23.23	29.43	31.00	32.87	30.55	29.66	29.75	29.55	28.69	28.37

Industry											
IT & Information Services (ISIC Rev.4 Code: 62, 63)	22.28	19.71	18.80	17.06	17.36	15.27	15.85	16.63	13.82	12.78	12.14
Financial Service Activities (ISIC Rev.4 Code: 64)	14.18	10.27	12.27	10.51	9.03	7.28	6.65	7.49	3.80	14.00	19.91
Insurance & Pensions (ISIC Rev.4 Code: 65)	10.74	2.03	7.25	16.04	15.76	14.82	14.75	9.74	6.23	3.13	3.06
Auxiliary Financial Services (ISIC Rev.4 Code: 66)	20.25	19.96	17.68	22.29	21.48	21.99	18.98	18.33	15.18	16.80	14.33
Legal, Accounting & Consultancy (ISIC Rev.4 Code: 69,70)							14.56	14.96	14.74	11.88	11.68
Architectural & Technical Services (ISIC Rev.4 Code: 71)							12.59	12.17	12.90	10.14	8.33
Research & Development (ISIC Rev.4 Code: 72)	7.84	5.84	5.41	6.25	5.97	5.02	5.09	5.48	5.99	4.53	4.25
Advertising & Market Research (ISIC Rev.4 Code: 73)							11.58	12.94	12.06	11.04	10.22
Other Professionals Services (ISIC Rev.4 Code: 74, 75)							24.31	26.19	25.92	27.30	26.34
Rental & Leasing Activities (ISIC Rev.4 Code: 77)							40.51	41.38	40.67	40.37	41.39
Employment Activities (ISIC Rev.4 Code: 78)							1.55	1.43	2.56	-6.86	1.53
Travel Service (ISIC Rev.4 Code: 79)							5.37	4.54	7.41	8.56	9.48
Security & Building Services (ISIC Rev.4 Code: 80, 81, 82)							9.27	10.11	9.93	9.71	10.40
Public Administration & Defense (ISIC Rev.4 Code: 84)	11.63	12.15	12.46	12.89	12.97	13.15	13.81	14.09	14.59	14.18	14.11

(continued)

Table 4.36 (continued)

Industry	2000	2001	2002	2003	2004	2005	2006	2007	2008	2009	2010
Education (ISIC Rev.4 Code: 85)	9.83	9.80	9.99	10.06	10.82	10.91	11.05	11.33	11.62	11.74	11.88
Human Health Activities (ISIC Rev.4 Code: 86)	27.01	27.14	27.62	27.57	27.82	27.66	27.65	28.49	28.91	29.00	28.54
Residential Care & Social Work (ISIC Rev.4 Code: 87, 88)	8.23	7.22	6.07	5.23	5.47	5.58	6.55	6.51	6.58	6.64	7.03
Creative Entertainment Activities (ISIC Rev.4 Code: 90, 91, 92)							20.45	19.56	18.59	18.12	16.92
Sports & Recreation Activities (ISIC Rev.4 Code: 93)							19.47	20.42	19.92	17.95	18.51
Membership Organizations (ISIC Rev.4 Code: 94)	3.21	0.63	0.08	0.43	−0.70	4.50	4.44	4.24	4.40	4.40	4.08
Repair Services (ISIC Rev.4 Code: 95)							19.03	20.73	20.90	17.23	14.71
Personal Service Activities (ISIC Rev.4 Code: 96)	43.35	42.47	43.14	42.62	42.75	43.23	42.64	43.77	42.38	41.20	39.99
Undifferentiated Services (ISIC Rev.4 Code: 97, 98)											
All Services (Average)	22.85	22.74	22.49	22.90	22.92	22.81	22.90	23.29	23.34	22.67	22.65
Business Sector Services (Average)	25.47	25.22	24.88	25.43	25.34	25.19	25.15	25.53	25.53	24.86	24.89
Business Sector Services (Average excluding real estate)	17.03	16.55	16.51	16.99	16.56	16.12	15.83	15.97	15.79	15.42	15.93
All Industries (Average)	18.92	18.91	18.84	19.08	19.04	18.79	18.81	19.06	18.86	18.61	18.65

Note: Gross Profit Ratio is computed by dividing gross surplus by gross productive output in the corresponding service sector. Gross surplus and gross productive output data are secured from OECD databases.

Table 4.37 Gross profit ratio: Germany

Industry	2000	2001	2002	2003	2004	2005	2006	2007	2008	2009	2010
Construction (ISIC Rev.4 Code: 41)	20.06	20.26	20.98	20.81	21.30	21.44	21.80	21.92	21.23	20.55	20.93
Automotive Retail & Repair (ISIC Rev.4 Code: 45)	17.94	19.18	19.40	19.51	19.70	21.17	21.49	21.87	20.54	19.39	20.65
Wholesale Trade (ISIC Rev.4 Code: 46)	20.86	23.01	26.93	26.57	26.93	28.34	29.16	24.52	21.65	22.58	20.45
Retail Trade (ISIC Rev.4 Code: 47)	12.92	15.98	16.38	16.97	18.44	22.34	21.85	25.44	25.74	22.07	24.78
Land Transport (ISIC Rev.4 Code: 49)	15.30	15.71	16.38	15.95	16.08	16.17	16.67	16.61	15.10	13.17	12.15
Water Transport (ISIC Rev.4 Code: 50)	20.87	22.43	22.19	20.44	18.73	19.38	21.38	21.01	18.97	16.70	16.49
Air Transport (ISIC Rev.4 Code: 51)	28.97	31.39	28.02	25.90	28.99	30.69	20.89	21.57	24.51	28.63	23.38
Support Activities for Transport (ISIC Rev.4 Code: 52)	17.26	10.36	8.38	5.41	7.37	6.81	8.69	10.71	4.18	1.28	2.25
Postal & Courier Activities (ISIC Rev.4 Code: 53)	8.16	9.71	13.35	14.34	15.68	15.71	17.41	16.80	15.32	13.55	13.21
Hotel & Restaurants (ISIC Rev.4 Code: 55, 56)	9.34	7.57	9.13	11.37	8.98	6.31	3.78	2.50	2.42	-2.29	-8.92
Publishing Activities (ISIC Rev.4 Code: 58)	17.10	16.92	14.75	14.66	15.02	17.06	16.61	18.54	21.19	24.48	23.45
Audiovisual & Broadcasting (ISIC Rev.4 Code: 59, 60)	13.72	11.57	10.43	10.65	11.81	12.20	12.81	15.66	21.57	25.32	24.41
Telecommunications (ISIC Rev.4 Code: 61)	20.98	22.90	19.75	19.31	18.75	22.64	21.03	22.29	20.64	23.27	22.15

(continued)

Table 4.37 (continued)

Industry	2000	2001	2002	2003	2004	2005	2006	2007	2008	2009	2010
IT & Information Services (ISIC Rev.4 Code: 62, 63)	37.19	37.20	40.44	29.51	35.19	28.66	32.49	28.38	26.07	24.97	24.19
Financial Service Activities (ISIC Rev.4 Code: 64)	11.03	11.52	10.38	6.15	5.91	7.06	7.81	10.02	6.21	10.76	7.23
Insurance & Pensions (ISIC Rev.4 Code: 65)	13.41	13.67	18.58	21.20	24.74	23.77	22.01	14.26	13.52	17.42	21.80
Auxiliary Financial Services (ISIC Rev.4 Code: 66)	3.91	-0.53	-1.90	2.50	5.94	-0.82	0.97	5.10	0.61	7.90	6.70
Legal, Accounting & Consultancy (ISIC Rev.4 Code: 69,70)	36.90	37.35	36.75	34.90	32.45	32.04	32.12	32.01	30.70	25.59	23.92
Architectural & Technical Services (ISIC Rev.4 Code: 71)	36.23	37.00	37.29	35.20	33.14	32.50	32.64	32.77	32.05	25.87	23.74
Research & Development (ISIC Rev.4 Code: 72)	38.36	38.20	35.42	34.06	30.47	30.76	30.62	29.87	27.04	24.95	24.29
Advertising & Market Research (ISIC Rev.4 Code: 73)	43.21	40.04	38.31	40.74	39.50	37.70	39.68	39.64	40.69	37.22	35.95
Other Professionals Services (ISIC Rev.4 Code: 74, 75)	44.18	39.59	35.93	38.46	36.71	33.83	38.04	39.56	39.40	33.02	33.04
Rental & Leasing Activities (ISIC Rev.4 Code: 77)	35.27	34.95	38.55	38.52	37.82	36.85	36.96	38.21	37.24	34.67	32.80
Employment Activities (ISIC Rev.4 Code: 78)	64.67	61.52	69.57	67.23	64.57	63.50	65.49	68.48	68.39	68.05	65.45
Travel Service (ISIC Rev.4 Code: 79)	17.36	18.61	16.35	17.89	12.95	5.31	10.50	10.87	12.00	14.09	16.26
Security & Building Services (ISIC Rev.4 Code: 80, 81, 82)	17.67	20.05	18.12	18.06	21.95	18.17	13.59	14.19	15.24	13.49	16.92

Public Administration & Defense (ISIC Rev.4 Code: 84)	12.91	12.75	15.13	15.60	16.80	20.31	21.00	21.38	20.76	17.01	15.46
Education (ISIC Rev.4 Code: 85)	11.56	11.51	11.24	11.22	11.27	11.19	11.35	11.77	11.55	11.21	11.09
Human Health Activities (ISIC Rev.4 Code: 86)	22.19	22.46	23.71	23.20	23.79	24.77	25.76	24.93	24.75	23.93	23.44
Residential Care & Social Work (ISIC Rev.4 Code: 87, 88)	29.50	29.62	30.96	30.27	30.89	31.73	32.74	31.54	31.31	30.31	30.13
Creative Entertainment Activities (ISIC Rev.4 Code: 90, 91, 92)	30.96	32.62	33.20	33.67	34.99	36.47	37.35	38.23	38.05	36.73	35.77
Sports & Recreation Activities (ISIC Rev.4 Code: 93)	31.40	33.84	34.28	34.47	35.55	37.28	38.55	39.45	39.07	39.62	39.26
Membership Organizations (ISIC Rev.4 Code: 94)	29.76	29.98	30.28	30.44	31.79	33.27	34.10	34.41	35.65	34.69	33.64
Repair Services (ISIC Rev.4 Code: 95)	2.01	1.92	2.44	2.43	2.76	2.88	2.93	3.08	3.39	3.18	3.17
Personal Service Activities (ISIC Rev.4 Code: 96)	22.88	19.50	9.27	6.06	9.00	14.75	16.07	11.21	15.94	6.40	9.76
Undifferentiated Services (ISIC Rev.4 Code: 97, 98)	49.01	49.60	50.98	51.62	53.08	54.33	55.22	55.43	56.85	55.75	54.97
All Services (Average)	26.98	27.20	28.07	27.62	27.99	28.04	28.48	28.70	28.01	27.06	26.59
Business Sector Services (Average)	30.55	30.83	31.94	31.36	31.56	31.23	31.63	31.81	30.86	30.10	29.61
Business Sector Services (Average excluding real estate)	22.09	22.27	22.90	22.48	22.96	22.62	22.97	22.84	21.40	20.56	20.43
All Industries (Average)	20.06	20.26	20.98	20.81	21.30	21.44	21.80	21.92	21.23	20.55	20.93

Note: Gross Profit Ratio is computed by dividing gross surplus by gross productive output in the corresponding service sector. Gross surplus and gross productive output data are secured from OECD databases.

Table 4.38 Gross profit ratio: Hungary

Industry	2000	2001	2002	2003	2004	2005	2006	2007	2008	2009	2010
Construction (ISIC Rev.4 Code: 41)	22.15	22.58	23.80	21.77	19.83	18.81	16.76	16.91	18.01	17.27	17.38
Automotive Retail & Repair (ISIC Rev.4 Code: 45)	14.40	19.80	21.46	17.36	12.16	11.47	12.36	14.91	17.15	11.10	11.70
Wholesale Trade (ISIC Rev.4 Code: 46)	10.11	14.95	18.02	15.91	16.44	17.13	19.26	18.12	18.61	14.95	16.19
Retail Trade (ISIC Rev.4 Code: 47)	11.70	16.52	17.60	13.22	13.78	12.88	12.53	12.62	11.12	6.74	7.79
Land Transport (ISIC Rev.4 Code: 49)	31.88	33.34	20.72	18.15	17.68	14.61	15.62	17.58	17.27	17.11	16.87
Water Transport (ISIC Rev.4 Code: 50)	–9.70	–2.24	2.18	0.42	0.79	11.28	5.69	9.38	13.08	1.81	4.19
Air Transport (ISIC Rev.4 Code: 51)	–8.38	–10.11	–1.60	–2.10	–7.60	–11.00	–10.06	–6.24	–7.80	–9.02	–4.37
Support Activities for Transport (ISIC Rev.4 Code: 52)	–13.02	–16.83	18.59	23.19	20.60	26.90	31.16	35.72	22.60	22.77	25.67
Postal & Courier Activities (ISIC Rev.4 Code: 53)	7.81	24.52	25.12	17.55	11.61	17.77	17.72	14.90	14.99	13.04	11.78
Hotel & Restaurants (ISIC Rev.4 Code: 55, 56)	16.50	14.82	15.41	12.82	11.75	11.08	9.67	9.13	8.27	8.48	6.50
Publishing Activities (ISIC Rev.4 Code: 58)	7.98	7.60	13.06	11.09	10.85	13.53	16.55	16.06	14.81	11.76	14.73
Audiovisual & Broadcasting (ISIC Rev.4 Code: 59, 60)	12.94	2.98	5.66	21.28	21.96	18.27	22.34	21.46	26.44	39.03	40.96
Telecommunications (ISIC Rev.4 Code: 61)	40.15	37.96	42.44	42.87	45.14	43.13	41.15	39.66	43.52	41.51	41.41

IT & Information Services (ISIC Rev.4 Code: 62, 63)	18.31	16.90	19.17	15.90	15.59	14.56	17.64	18.65	18.77	18.36	19.65
Financial Service Activities (ISIC Rev.4 Code: 64)	22.86	25.55	28.47	30.58	28.72	30.66	30.48	22.14	20.03	23.42	25.29
Insurance & Pensions (ISIC Rev.4 Code: 65)	18.29	17.14	18.64	17.50	27.98	23.30	21.44	10.75	6.45	18.24	11.68
Auxiliary Financial Services (ISIC Rev.4 Code: 66)	41.77	40.94	39.92	41.14	38.48	41.78	50.14	48.26	46.35	50.16	53.02
Legal, Accounting & Consultancy (ISIC Rev.4 Code: 69,70)	26.61	25.74	31.17	27.67	27.45	19.50	24.05	21.19	21.05	20.44	18.23
Architectural & Technical Services (ISIC Rev.4 Code: 71)	22.15	23.54	28.08	19.73	20.71	17.93	19.73	20.74	22.69	23.16	25.23
Research & Development (ISIC Rev.4 Code: 72)	23.23	22.24	9.93	10.22	12.87	16.39	17.76	8.08	6.90	7.71	6.35
Advertising & Market Research (ISIC Rev.4 Code: 73)	17.65	18.43	20.33	13.11	16.21	16.13	18.23	18.05	17.94	16.03	16.43
Other Professionals Services (ISIC Rev.4 Code: 74, 75)	37.32	36.74	39.48	30.49	29.26	30.03	32.38	21.90	31.63	31.32	27.86
Rental & Leasing Activities (ISIC Rev.4 Code: 77)	48.52	50.25	52.88	51.96	55.36	55.53	56.69	51.98	52.40	52.66	50.11
Employment Activities (ISIC Rev.4 Code: 78)	-17.91	-10.16	-9.50	2.49	-14.94	0.73	4.18	2.45	-0.29	2.59	1.65
Travel Service (ISIC Rev.4 Code: 79)	0.12	-2.19	12.13	5.93	16.25	7.33	4.93	6.30	6.76	9.71	9.72
Security & Building Services (ISIC Rev.4 Code: 80, 81, 82)	30.42	30.13	31.72	31.68	27.23	27.71	27.60	24.99	27.20	27.20	19.22
Public Administration & Defense (ISIC Rev.4 Code: 84)	26.81	25.31	22.83	22.31	21.86	21.17	20.07	20.85	21.28	22.13	22.42

(continued)

Table 4.38 (continued)

Industry	2000	2001	2002	2003	2004	2005	2006	2007	2008	2009	2010
Education (ISIC Rev.4 Code: 85)	16.66	15.53	14.38	12.01	12.17	12.37	12.94	13.93	15.56	16.51	17.10
Human Health Activities (ISIC Rev.4 Code: 86)	16.52	15.24	13.18	13.28	11.40	11.83	11.75	10.68	10.75	10.31	11.34
Residential Care & Social Work (ISIC Rev.4 Code: 87, 88)	3.81	3.22	3.58	2.83	4.60	2.48	2.40	2.27	5.73	1.81	2.77
Creative Entertainment Activities (ISIC Rev.4 Code: 90, 91, 92)	28.54	32.32	34.02	24.06	15.15	22.10	20.14	20.99	18.95	15.21	17.52
Sports & Recreation Activities (ISIC Rev.4 Code: 93)	9.34	9.95	8.53	5.96	7.28	7.75	9.61	10.42	11.55	3.98	7.39
Membership Organizations (ISIC Rev.4 Code: 94)	18.62	18.00	17.87	20.52	15.77	15.41	16.23	16.04	17.47	17.54	18.00
Repair Services (ISIC Rev.4 Code: 95)	17.40	17.03	20.16	11.06	11.50	11.60	10.59	8.70	21.14	18.89	17.22
Personal Service Activities (ISIC Rev.4 Code: 96)	40.46	40.61	30.82	35.39	38.51	35.51	34.51	36.08	37.47	36.29	34.73
Undifferentiated Services (ISIC Rev.4 Code: 97, 98)											1.65
All Services (Average)	24.47	25.01	25.61	24.02	23.47	23.03	23.49	22.85	22.89	22.96	23.27
Business Sector Services (Average)	25.56	26.57	28.22	26.67	26.12	25.49	26.15	25.00	24.68	24.96	25.13
Business Sector Services (Average excluding real estate)	19.20	20.64	22.60	20.94	20.55	20.04	20.99	19.61	19.26	18.96	19.06
All Industries (Average)	18.47	19.10	20.09	19.36	19.75	19.26	19.14	18.63	18.69	19.23	19.84

Note: Gross Profit Ratio is computed by dividing gross surplus by gross productive output in the corresponding service sector. Gross surplus and gross productive output data are secured from OECD databases.

Table 4.39 Gross profit ratio: Italy

Industry	2000	2001	2002	2003	2004	2005	2006	2007	2008	2009	2010
Construction (ISIC Rev.4 Code: 41)	18.73	19.71	19.72	20.43	21.28	21.04	20.89	20.67	20.63	20.77	20.47
Automotive Retail & Repair (ISIC Rev.4 Code: 45)	24.09	27.21	24.48	23.61	22.61	21.74	20.44	19.18	19.24	19.02	18.57
Wholesale Trade (ISIC Rev.4 Code: 46)	31.90	32.23	31.91	31.16	31.84	30.75	29.57	28.28	27.48	26.11	26.20
Retail Trade (ISIC Rev.4 Code: 47)	43.24	41.37	37.96	37.60	36.69	35.19	33.37	33.21	31.89	29.61	29.63
Land Transport (ISIC Rev.4 Code: 49)	21.07	21.04	22.32	21.61	22.67	20.58	19.73	21.09	20.95	22.46	22.29
Water Transport (ISIC Rev.4 Code: 50)	24.07	21.49	21.81	22.59	24.13	23.03	19.98	21.38	20.52	15.77	12.44
Air Transport (ISIC Rev.4 Code: 51)	4.23	6.98	6.21	8.32	9.64	6.16	3.79	1.59	1.76	3.37	2.29
Support Activities for Transport (ISIC Rev.4 Code: 52)	13.89	15.82	16.08	14.18	13.78	13.56	12.52	11.91	11.20	14.14	14.45
Postal & Courier Activities (ISIC Rev.4 Code: 53)	−1.53	4.82	3.17	2.93	2.80	2.92	2.11	10.37	4.89	3.96	6.78
Hotel & Restaurants (ISIC Rev.4 Code: 55, 56)	31.83	31.97	30.01	28.64	28.63	28.86	28.77	29.06	27.83	28.31	27.21
Publishing Activities (ISIC Rev.4 Code: 58)	13.08	14.39	17.48	14.97	13.99	13.30	13.08	13.45	12.81	11.95	11.72
Audiovisual & Broadcasting (ISIC Rev.4 Code: 59, 60)	37.62	35.67	35.46	36.70	36.97	36.21	35.29	37.13	38.48	39.27	39.18
Telecommunications (ISIC Rev.4 Code: 61)	32.89	37.15	41.31	41.00	41.61	40.85	39.28	38.41	38.18	36.34	37.16

(continued)

Table 4.39 (continued)

Industry	2000	2001	2002	2003	2004	2005	2006	2007	2008	2009	2010
IT & Information Services (ISIC Rev.4 Code: 62, 63)	23.09	21.66	22.39	22.07	20.95	20.53	18.75	19.23	19.90	18.79	19.59
Financial Service Activities (ISIC Rev.4 Code: 64)	28.12	22.92	22.98	24.49	25.29	23.15	20.30	23.69	26.26	30.25	31.52
Insurance & Pensions (ISIC Rev.4 Code: 65)	−7.33	25.80	23.84	28.17	18.57	25.95	21.47	29.90	23.89	1.69	9.53
Auxiliary Financial Services (ISIC Rev.4 Code: 66)	30.11	28.86	26.43	28.00	29.54	28.63	24.29	24.44	23.55	20.31	14.94
Legal, Accounting & Consultancy (ISIC Rev.4 Code: 69,70)	60.15	55.56	54.58	56.85	55.54	53.99	51.12	50.09	50.51	49.59	48.39
Architectural & Technical Services (ISIC Rev.4 Code: 71)	41.34	38.27	37.59	38.36	37.70	36.66	35.91	35.11	35.46	33.86	32.82
Research & Development (ISIC Rev.4 Code: 72)	16.11	14.70	14.06	14.34	13.15	12.70	12.99	12.56	10.75	10.40	10.58
Advertising & Market Research (ISIC Rev.4 Code: 73)	9.57	8.55	8.63	9.44	9.37	8.92	8.44	7.77	7.35	6.12	5.59
Other Professionals Services (ISIC Rev.4 Code: 74, 75)	54.38	51.04	50.25	51.01	50.66	49.87	48.78	47.89	47.85	42.77	42.13
Rental & Leasing Activities (ISIC Rev.4 Code: 77)	42.51	42.41	40.69	41.44	40.00	39.66	37.23	37.27	37.42	37.19	36.49
Employment Activities (ISIC Rev.4 Code: 78)	13.62	9.47	8.64	11.03	9.28	7.02	6.48	4.91	4.24	−15.33	−32.95
Travel Service (ISIC Rev.4 Code: 79)	6.99	6.19	6.31	5.71	6.25	6.42	6.26	5.80	5.39	4.43	4.52
Security & Building Services (ISIC Rev.4 Code: 80, 81, 82)	17.06	15.49	15.15	15.97	14.90	14.74	13.60	12.95	12.20	13.07	12.53

Public Administration & Defense (ISIC Rev.4 Code: 84)	15.35	15.27	15.33	14.98	15.28	15.58	15.95	16.54	16.46	16.31	16.85
Education (ISIC Rev.4 Code: 85)	7.86	8.59	8.85	9.64	10.09	8.83	8.95	8.99	8.57	8.06	8.16
Human Health Activities (ISIC Rev.4 Code: 86)	17.02	16.92	16.70	16.99	16.52	16.65	15.62	16.17	15.77	16.95	17.16
Residential Care & Social Work (ISIC Rev.4 Code: 87, 88)	10.00	10.04	9.81	9.96	10.78	12.51	12.06	10.84	10.04	10.46	10.20
Creative Entertainment Activities (ISIC Rev.4 Code: 90, 91, 92)	27.35	28.41	29.09	29.55	28.09	29.48	29.66	30.64	30.60	29.52	29.74
Sports & Recreation Activities (ISIC Rev.4 Code: 93)	31.53	29.66	29.40	29.43	30.50	29.01	27.89	28.98	29.72	27.33	27.64
Membership Organizations (ISIC Rev.4 Code: 94)	7.03	5.27	5.18	4.29	4.86	4.47	4.85	4.78	4.21	3.18	2.95
Repair Services (ISIC Rev.4 Code: 95)	44.40	42.63	43.60	44.29	44.15	42.75	41.11	40.42	39.43	36.95	36.02
Personal Service Activities (ISIC Rev.4 Code: 96)	50.89	51.69	45.87	44.36	44.92	43.95	41.97	39.31	39.94	40.68	40.30
Undifferentiated Services (ISIC Rev.4 Code: 97, 98)											
All Services (Average)	32.64	32.51	32.37	32.48	32.52	31.90	31.05	31.18	31.17	31.02	30.92
Business Sector Services (Average)	37.32	37.10	37.07	37.29	37.33	36.61	35.57	35.61	35.76	35.85	35.61
Business Sector Services (Average excluding real estate)	29.82	29.57	29.09	29.12	28.91	27.93	26.36	26.54	26.20	25.85	25.61
All Industries (Average)	23.94	24.06	24.04	24.09	24.11	23.38	22.60	22.53	22.34	23.38	22.96

Note: Gross Profit Ratio is computed by dividing gross surplus by gross productive output in the corresponding service sector. Gross surplus and gross productive output data are secured from OECD databases.

Table 4.40 Gross profit ratio: Korea

Industry	2000	2001	2002	2003	2004	2005	2006	2007	2008	2009	2010
Construction (ISIC Rev.4 Code: 41)	12.48	12.13	11.76	12.05	11.25	11.69	12.53	12.43	10.59	10.61	9.76
Automotive Retail & Repair (ISIC Rev.4 Code: 45)											
Wholesale Trade (ISIC Rev.4 Code: 46)											
Retail Trade (ISIC Rev.4 Code: 47)											
Land Transport (ISIC Rev.4 Code: 49)											
Water Transport (ISIC Rev.4 Code: 50)											
Air Transport (ISIC Rev.4 Code: 51)											
Support Activities for Transport (ISIC Rev.4 Code: 52)											
Postal & Courier Activities (ISIC Rev.4 Code: 53)											
Hotel & Restaurants (ISIC Rev.4 Code: 55, 56)	13.35	13.35	13.68	12.58	12.71	13.22	13.47	13.59	12.82	11.82	11.90
Publishing Activities (ISIC Rev.4 Code: 58)	13.55	12.89	12.26	12.06	11.91	11.96	12.59	14.11	12.92	12.94	11.01
Audiovisual & Broadcasting (ISIC Rev.4 Code: 59, 60)	11.32	11.17	11.11	11.22	11.16	11.43	12.24	12.24	11.81	14.30	13.73
Telecommunications (ISIC Rev.4 Code: 61)	25.45	34.24	36.13	32.24	30.85	30.06	28.16	25.63	24.62	23.92	26.29

IT & Information Services (ISIC Rev.4 Code: 62, 63)	23.72	23.35	24.50	24.84	24.64	24.87	24.71	28.32	21.54	16.79	14.45
Financial Service Activities (ISIC Rev.4 Code: 64)											
Insurance & Pensions (ISIC Rev.4 Code: 65)											
Auxiliary Financial Services (ISIC Rev.4 Code: 66)											
Legal, Accounting & Consultancy (ISIC Rev.4 Code: 69,70)											
Architectural & Technical Services (ISIC Rev.4 Code: 71)											
Research & Development (ISIC Rev.4 Code: 72)	9.46	9.80	9.72	9.67	9.70	9.63	9.69	9.50	9.91	10.15	10.06
Advertising & Market Research (ISIC Rev.4 Code: 73)	2.76	2.44	2.32	2.18	2.15	2.13	2.01	2.00	1.75	1.80	2.37
Other Professionals Services (ISIC Rev.4 Code: 74, 75)											
Rental & Leasing Activities (ISIC Rev.4 Code: 77)	47.75	45.64	46.40	45.86	46.16	46.05	47.05	45.92	43.61	43.52	43.36
Employment Activities (ISIC Rev.4 Code: 78)											
Travel Service (ISIC Rev.4 Code: 79)											
Security & Building Services (ISIC Rev.4 Code: 80, 81, 82)											
Public Administration & Defense (ISIC Rev.4 Code: 84)	19.25	19.43	19.66	20.02	20.11	19.61	20.23	20.46	20.82	21.29	22.45

(continued)

Table 4.40 (continued)

Industry	2000	2001	2002	2003	2004	2005	2006	2007	2008	2009	2010
Education (ISIC Rev.4 Code: 85)	11.42	11.36	12.00	12.28	12.33	12.43	12.66	12.70	12.80	12.83	13.05
Human Health Activities (ISIC Rev.4 Code: 86)	15.42	16.89	15.76	15.97	16.12	16.18	16.30	17.08	16.39	16.81	16.77
Residential Care & Social Work (ISIC Rev.4 Code: 87, 88)	3.56	3.59	3.65	3.71	3.72	3.70	3.71	3.77	3.85	4.03	4.08
Creative Entertainment Activities (ISIC Rev.4 Code: 90, 91, 92)	12.79	13.29	12.86	13.29	13.48	13.45	13.19	12.98	13.31	13.78	14.51
Sports & Recreation Activities (ISIC Rev.4 Code: 93)	29.87	30.42	29.09	29.77	29.56	29.75	28.11	26.38	26.38	27.07	26.48
Membership Organizations (ISIC Rev.4 Code: 94)	9.19	9.19	9.26	9.22	9.20	9.20	9.16	9.10	9.03	9.07	9.02
Repair Services (ISIC Rev.4 Code: 95)											
Personal Service Activities (ISIC Rev.4 Code: 96)	27.79	27.29	27.18	27.14	27.32	27.41	27.42	26.96	25.16	26.97	26.93
Undifferentiated Services (ISIC Rev.4 Code: 97, 98)											
All Services (Average)	27.89	28.09	28.55	28.13	27.95	27.52	26.97	26.67	25.85	25.75	25.96
Business Sector Services (Average)	31.55	31.92	32.53	32.07	31.94	31.51	30.82	30.38	29.29	29.21	29.23
Business Sector Services (Average excluding real estate)	24.70	25.54	26.67	26.09	26.00	25.57	24.98	24.80	23.88	23.46	23.93
All Industries (Average)	20.88	20.73	20.96	20.50	19.99	19.05	18.50	18.22	16.55	17.17	17.30

Note: Gross Profit Ratio is computed by dividing gross surplus by gross productive output in the corresponding service sector. Gross surplus and gross productive output data are secured from OECD databases.

Table 4.41 Gross profit ratio: Netherlands

Industry	2000	2001	2002	2003	2004	2005	2006	2007	2008	2009	2010
Construction (ISIC Rev.4 Code: 41)	11.17	11.51	12.13	12.45	12.26	13.07	13.70	14.05	14.33	14.57	13.21
Automotive Retail & Repair (ISIC Rev.4 Code: 45)	18.90	19.71	20.88	20.24	19.46	18.75	19.05	19.05	17.80	14.80	18.01
Wholesale Trade (ISIC Rev.4 Code: 46)	26.87	25.77	26.97	25.95	26.86	28.22	28.17	29.44	27.87	23.30	26.76
Retail Trade (ISIC Rev.4 Code: 47)	25.82	24.76	24.47	23.12	20.20	19.71	19.52	18.54	15.68	16.53	13.56
Land Transport (ISIC Rev.4 Code: 49)	21.41	21.14	20.39	19.43	18.23	18.39	18.97	19.14	17.48	17.72	16.67
Water Transport (ISIC Rev.4 Code: 50)	22.36	23.82	20.79	21.00	22.95	23.17	24.02	22.16	20.61	15.05	11.73
Air Transport (ISIC Rev.4 Code: 51)	8.37	5.80	8.14	0.62	0.55	−0.79	−0.85	−4.22	−8.26	−20.57	−18.74
Support Activities for Transport (ISIC Rev.4 Code: 52)	28.45	28.67	27.30	27.59	28.17	30.31	30.81	31.00	31.70	28.97	30.63
Postal & Courier Activities (ISIC Rev.4 Code: 53)	17.37	17.47	15.57	16.96	21.89	23.33	22.59	22.92	22.43	21.18	18.50
Hotel & Restaurants (ISIC Rev.4 Code: 55, 56)	23.47	22.74	24.79	24.27	24.49	24.41	23.73	24.52	22.74	22.37	21.74
Publishing Activities (ISIC Rev.4 Code: 58)	21.11	21.07	23.81	23.39	25.04	23.60	23.63	23.90	22.93	24.00	27.01
Audiovisual & Broadcasting (ISIC Rev.4 Code: 59, 60)	15.57	15.80	16.97	17.64	18.69	17.65	17.98	17.91	17.46	17.59	19.57
Telecommunications (ISIC Rev.4 Code: 61)	24.34	23.18	31.25	36.66	36.09	36.63	35.67	34.80	33.09	32.17	33.00

(continued)

Table 4.41 (continued)

Industry	2000	2001	2002	2003	2004	2005	2006	2007	2008	2009	2010
IT & Information Services (ISIC Rev.4 Code: 62, 63)	15.88	15.49	13.76	13.09	14.04	14.23	15.13	16.56	15.43	11.22	12.04
Financial Service Activities (ISIC Rev.4 Code: 64)	16.36	16.52	21.80	24.37	25.01	25.44	14.75	6.21	13.01	24.75	32.30
Insurance & Pensions (ISIC Rev.4 Code: 65)	22.12	24.52	21.40	31.03	31.03	37.23	36.12	32.75	24.70	27.02	24.30
Auxiliary Financial Services (ISIC Rev.4 Code: 66)	30.61	28.47	24.64	25.26	25.21	22.51	26.76	27.75	18.16	16.35	18.40
Legal, Accounting & Consultancy (ISIC Rev.4 Code: 69,70)	14.81	14.78	13.11	12.44	12.58	13.57	13.82	14.54	13.36	11.37	11.66
Architectural & Technical Services (ISIC Rev.4 Code: 71)	9.04	8.03	8.55	8.43	8.51	10.01	11.58	11.91	12.91	11.04	8.86
Research & Development (ISIC Rev.4 Code: 72)	3.76	4.97	4.84	4.12	7.67	9.92	11.45	8.47	5.66	2.86	9.54
Advertising & Market Research (ISIC Rev.4 Code: 73)	17.49	15.99	15.54	14.24	14.44	15.44	15.96	16.66	15.65	12.40	12.03
Other Professionals Services (ISIC Rev.4 Code: 74, 75)	23.81	25.42	24.52	24.00	25.36	25.27	24.96	26.10	27.91	28.07	26.26
Rental & Leasing Activities (ISIC Rev.4 Code: 77)	44.48	43.93	40.60	40.65	40.20	40.71	41.87	42.45	38.94	37.44	35.32
Employment Activities (ISIC Rev.4 Code: 78)	8.89	10.82	10.22	9.53	8.48	10.26	13.11	15.02	16.44	16.30	15.13
Travel Service (ISIC Rev.4 Code: 79)	3.00	4.93	4.87	6.82	6.52	6.52	6.44	7.80	6.68	13.01	13.44
Security & Building Services (ISIC Rev.4 Code: 80, 81, 82)	12.74	15.71	15.44	15.03	14.94	14.49	14.04	13.89	13.90	14.31	13.15

Public Administration & Defense (ISIC Rev.4 Code: 84)	16.68	15.92	15.70	15.63	15.96	16.17	16.12	16.30	16.17	15.68	16.03
Education (ISIC Rev.4 Code: 85)	11.19	11.13	11.38	11.06	11.00	11.18	11.25	11.34	11.42	11.52	11.70
Human Health Activities (ISIC Rev.4 Code: 86)	23.48	24.45	26.23	25.03	25.15	23.82	23.53	25.82	26.33	27.56	28.47
Residential Care & Social Work (ISIC Rev.4 Code: 87, 88)	11.03	10.05	11.76	11.42	10.82	11.36	10.88	10.01	10.16	11.98	9.64
Creative Entertainment Activities (ISIC Rev.4 Code: 90, 91, 92)	17.29	17.26	17.63	17.32	17.86	17.81	16.89	16.69	16.23	16.68	17.16
Sports & Recreation Activities (ISIC Rev.4 Code: 93)	6.36	3.92	5.46	3.71	4.91	4.89	4.32	4.40	3.44	4.52	4.61
Membership Organizations (ISIC Rev.4 Code: 94)	3.51	4.15	4.51	4.63	4.64	5.31	5.08	3.56	3.31	5.57	4.37
Repair Services (ISIC Rev.4 Code: 95)	25.14	26.54	26.96	27.70	25.82	24.42	24.10	25.42	25.92	26.07	25.94
Personal Service Activities (ISIC Rev.4 Code: 96)	34.35	35.17	35.66	34.39	32.59	33.20	33.04	33.45	33.06	32.99	32.88
Undifferentiated Services (ISIC Rev.4 Code: 97, 98)	97.99	98.09	98.02	98.06	98.07	98.12	98.13	98.13	98.15	98.13	97.92
All Services (Average)	23.46	23.08	23.31	23.09	23.26	23.70	23.50	23.63	22.73	21.21	21.96
Business Sector Services (Average)	25.64	25.27	25.47	25.44	25.62	26.22	26.00	26.05	24.85	22.69	23.74
Business Sector Services (Average excluding real estate)	20.49	20.33	21.07	21.54	21.61	22.31	21.43	20.91	19.65	19.34	20.82
All Industries (Average)	19.94	19.86	20.08	20.19	20.24	20.75	20.77	20.83	20.19	19.34	19.74

Note: Gross Profit Ratio is computed by dividing gross surplus by gross productive output in the corresponding service sector. Gross surplus and gross productive output data are secured from OECD databases.

Table 4.42 Gross profit ratio: Norway

Industry	2000	2001	2002	2003	2004	2005	2006	2007	2008	2009	2010
Construction (ISIC Rev.4 Code: 41)	10.00	10.21	11.11	10.36	12.77	13.22	13.86	13.49	12.18	13.15	14.36
Automotive Retail & Repair (ISIC Rev.4 Code: 45)	15.26	11.22	11.76	12.42	6.12	6.36	-1.17	18.54	20.74	16.43	
Wholesale Trade (ISIC Rev.4 Code: 46)	17.88	20.84	15.78	14.59	15.56	14.71	17.65	18.47	18.96	15.40	
Retail Trade (ISIC Rev.4 Code: 47)	7.97	7.40	10.12	8.26	8.09	8.86	10.32	12.19	1.92	3.95	
Land Transport (ISIC Rev.4 Code: 49)	32.01	36.39	34.15	34.85	35.56	36.44	39.38	35.26	36.21	35.39	33.68
Water Transport (ISIC Rev.4 Code: 50)	16.03	18.74	14.46	14.03	14.69	15.50	15.65	14.55	13.54	11.21	6.40
Air Transport (ISIC Rev.4 Code: 51)	6.85	-0.46	6.84	4.33	4.67	6.68	8.76	-0.63	-1.31	-0.41	
Support Activities for Transport (ISIC Rev.4 Code: 52)	14.80	15.72	13.11	12.34	13.14	13.95	12.99	13.26	9.84	9.32	
Postal & Courier Activities (ISIC Rev.4 Code: 53)	9.03	11.01	6.72	10.42	15.81	16.44	10.63	5.78	4.99	20.54	19.40
Hotel & Restaurants (ISIC Rev.4 Code: 55, 56)	9.42	9.74	11.28	8.79	11.19	12.05	14.40	13.18	12.57	12.63	12.33
Publishing Activities (ISIC Rev.4 Code: 58)	8.69	7.24	9.47	13.17	15.77	16.00	13.62	13.65	9.14	11.83	
Audiovisual & Broadcasting (ISIC Rev.4 Code: 59, 60)	12.82	11.15	8.02	12.78	14.15	19.16	16.99	13.92	19.60	14.27	
Telecommunications (ISIC Rev.4 Code: 61)	14.17	17.65	22.80	24.60	24.43	25.39	26.34	21.23	14.57	18.82	17.15

IT & Information Services (ISIC Rev.4 Code: 62, 63)	7.52	3.95	6.13	10.10	16.32	16.86	10.22	13.49	14.53	13.30	11.51
Financial Service Activities (ISIC Rev.4 Code: 64)	36.65	35.87	35.26	42.65	45.77	44.72	39.47	39.16	41.62	45.97	
Insurance & Pensions (ISIC Rev.4 Code: 65)	-8.44	-10.76	4.30	28.85	44.04	45.16	43.07	52.46	50.55	46.86	
Auxiliary Financial Services (ISIC Rev.4 Code: 66)	28.63	16.95	-6.90	12.17	7.74	-13.18	-8.23	-4.85	-18.94	-4.36	
Legal, Accounting & Consultancy (ISIC Rev.4 Code: 69,70)	21.54	25.60	25.05	26.57	26.76	28.02	20.84	20.48	25.81	25.95	
Architectural & Technical Services (ISIC Rev.4 Code: 71)	3.77	8.82	8.56	9.98	10.78	12.89	11.89	14.21	14.54	11.93	
Research & Development (ISIC Rev.4 Code: 72)	14.56	5.30	2.52	5.48	8.06	-1.97	4.55	3.15	4.31	8.93	13.76
Advertising & Market Research (ISIC Rev.4 Code: 73)	6.45	4.24	4.25	6.67	7.34	8.71	12.36	9.16	4.23	5.77	
Other Professionals Services (ISIC Rev.4 Code: 74, 75)	9.21	11.53	11.42	14.90	14.95	20.26	14.96	18.03	16.49	18.09	
Rental & Leasing Activities (ISIC Rev.4 Code: 77)	21.92	22.37	23.55	22.88	23.62	24.52	32.51	39.85	30.86	30.08	
Employment Activities (ISIC Rev.4 Code: 78)	9.91	10.27	9.71	6.79	6.62	2.44	-2.69	1.18	0.01	3.08	
Travel Service (ISIC Rev.4 Code: 79)	6.33	6.06	1.82	2.75	2.25	4.18	5.37	8.06	5.76	4.55	
Security & Building Services (ISIC Rev.4 Code: 80, 81, 82)	10.35	11.36	11.91	13.41	13.35	16.85	11.00	11.62	8.58	11.30	
Public Administration & Defense (ISIC Rev.4 Code: 84)	12.10	11.60	11.44	12.05	12.31	12.68	12.79	12.64	12.88	12.84	13.17

(continued)

Table 4.42 (continued)

Industry	2000	2001	2002	2003	2004	2005	2006	2007	2008	2009	2010
Education (ISIC Rev.4 Code: 85)	6.79	6.43	6.36	6.00	6.57	7.13	6.89	8.35	8.21	8.14	8.57
Human Health Activities (ISIC Rev.4 Code: 86)	16.07	15.91	16.05	16.17	15.01	15.32	15.08	14.82	15.63	15.11	14.90
Residential Care & Social Work (ISIC Rev.4 Code: 87, 88)	7.11	6.84	6.09	7.17	6.92	7.47	6.90	6.56	6.43	6.60	6.49
Creative Entertainment Activities (ISIC Rev.4 Code: 90, 91, 92)	42.29	47.52	47.97	42.85	40.95	38.73	37.25	35.51	32.87	31.39	
Sports & Recreation Activities (ISIC Rev.4 Code: 93)	20.62	21.52	21.98	20.70	21.07	19.62	20.64	17.94	15.51	11.46	
Membership Organizations (ISIC Rev.4 Code: 94)	1.26	1.39	1.17	0.85	2.54	2.95	4.33	5.02	4.91	5.05	
Repair Services (ISIC Rev.4 Code: 95)	11.64	5.09	6.68	15.12	16.71	12.21	10.58	16.76	-1.24	15.09	
Personal Service Activities (ISIC Rev.4 Code: 96)	25.92	26.51	28.20	29.08	32.51	31.91	35.04	38.10	36.96	35.41	
Undifferentiated Services (ISIC Rev.4 Code: 97, 98)											1.96
All Services (Average)	19.57	20.09	19.87	20.72	21.30	21.45	20.91	21.04	19.82	20.13	20.00
Business Sector Services (Average)	22.41	23.30	23.14	24.30	25.02	25.05	24.25	24.41	22.78	23.67	23.46
Business Sector Services (Average excluding real estate)	15.49	16.32	15.62	17.16	18.33	18.52	17.60	18.21	16.69	17.16	16.79
All Industries (Average)	27.75	27.05	25.64	26.36	27.96	29.74	29.82	27.46	28.49	25.00	26.12

Note: Gross Profit Ratio is computed by dividing gross surplus by gross productive output in the corresponding service sector. Gross surplus and gross productive output data are secured from OECD databases.

Table 4.43 Gross profit ratio: Slovenia

Industry	2000	2001	2002	2003	2004	2005	2006	2007	2008	2009	2010
Construction (ISIC Rev.4 Code: 41)	14.41	14.50	13.42	13.44	12.72	13.71	14.03	14.03	12.94	12.98	9.90
Automotive Retail & Repair (ISIC Rev.4 Code: 45)	22.72	21.22	20.42	21.24	19.61	18.44	19.62	20.88	19.97	16.71	18.40
Wholesale Trade (ISIC Rev.4 Code: 46)	15.73	18.00	19.08	20.24	20.36	19.13	19.07	20.50	20.98	18.60	16.44
Retail Trade (ISIC Rev.4 Code: 47)	21.33	22.87	23.56	23.70	20.29	23.38	21.00	21.03	21.24	22.04	22.20
Land Transport (ISIC Rev.4 Code: 49)	23.68	22.98	21.43	23.50	22.49	19.07	19.23	18.96	17.59	16.85	15.53
Water Transport (ISIC Rev.4 Code: 50)	8.35	15.16	30.39	30.02	39.93	37.05	33.38	39.24	36.02	34.20	29.55
Air Transport (ISIC Rev.4 Code: 51)	8.15	12.61	11.75	11.28	8.79	5.77	6.92	10.36	9.60	3.19	−2.38
Support Activities for Transport (ISIC Rev.4 Code: 52)	21.59	21.92	21.25	21.13	21.41	25.69	24.38	25.22	25.35	26.85	28.80
Postal & Courier Activities (ISIC Rev.4 Code: 53)	13.73	15.65	15.73	15.98	14.69	19.86	21.09	17.60	16.96	16.53	16.09
Hotel & Restaurants (ISIC Rev.4 Code: 55, 56)	21.02	21.26	20.32	20.25	17.88	17.77	18.13	18.59	17.26	15.17	14.20
Publishing Activities (ISIC Rev.4 Code: 58)	10.27	10.08	8.46	8.24	7.99	8.67	8.60	8.09	5.84	6.31	2.93
Audiovisual & Broadcasting (ISIC Rev.4 Code: 59, 60)	16.69	19.08	14.43	16.64	19.15	16.71	16.95	15.48	15.07	11.31	12.32
Telecommunications (ISIC Rev.4 Code: 61)	30.76	29.21	30.43	32.79	33.03	32.02	30.47	31.38	27.61	23.70	25.20

(continued)

Table 4.43 (continued)

Industry	2000	2001	2002	2003	2004	2005	2006	2007	2008	2009	2010
IT & Information Services (ISIC Rev.4 Code: 62, 63)	13.74	13.73	17.20	16.65	14.92	15.00	15.51	16.40	16.10	16.05	17.03
Financial Service Activities (ISIC Rev.4 Code: 64)	39.35	34.80	30.34	32.17	33.61	31.19	34.66	29.72	32.93	35.05	35.59
Insurance & Pensions (ISIC Rev.4 Code: 65)	3.60	−5.30	11.08	9.49	6.92	12.79	15.71	18.41	15.85	15.70	23.18
Auxiliary Financial Services (ISIC Rev.4 Code: 66)	25.37	22.68	9.85	15.06	15.29	12.28	14.68	21.40	6.17	6.77	15.10
Legal, Accounting & Consultancy (ISIC Rev.4 Code: 69,70)	28.06	30.04	29.09	29.67	31.51	29.43	29.49	28.17	27.89	26.15	26.24
Architectural & Technical Services (ISIC Rev.4 Code: 71)	9.35	8.67	9.27	12.01	10.71	11.32	11.87	13.89	14.31	13.97	13.69
Research & Development (ISIC Rev.4 Code: 72)	9.19	10.05	10.66	9.87	13.68	12.36	14.91	13.26	12.53	13.14	12.49
Advertising & Market Research (ISIC Rev.4 Code: 73)	−2.20	5.04	2.67	1.81	4.40	3.83	4.30	6.01	5.72	4.71	5.79
Other Professionals Services (ISIC Rev.4 Code: 74, 75)	19.03	21.20	36.67	37.06	38.81	39.68	37.04	41.25	42.61	42.78	41.75
Rental & Leasing Activities (ISIC Rev.4 Code: 77)	24.22	14.80	29.69	29.75	25.22	28.11	30.48	33.19	29.38	27.45	29.70
Employment Activities (ISIC Rev.4 Code: 78)	2.33	2.10	2.29	2.30	3.01	3.37	1.22	0.87	0.98	0.30	0.11
Travel Service (ISIC Rev.4 Code: 79)	8.34	6.03	4.26	3.75	3.68	2.58	2.01	4.12	1.81	1.51	1.98
Security & Building Services (ISIC Rev.4 Code: 80, 81, 82)	14.62	12.57	12.59	11.81	11.71	11.54	11.37	12.85	11.72	11.04	13.09

Public Administration & Defense (ISIC Rev.4 Code: 84)	10.61	10.30	9.63	10.44	11.44	11.86	12.32	13.45	13.52	14.21	15.09
Education (ISIC Rev.4 Code: 85)	5.73	5.43	5.14	5.02	4.93	5.15	5.20	5.57	5.47	5.51	5.66
Human Health Activities (ISIC Rev.4 Code: 86)	13.93	13.44	14.15	14.18	13.19	14.07	14.07	14.11	13.86	12.65	12.98
Residential Care & Social Work (ISIC Rev.4 Code: 87, 88)	5.03	4.30	3.91	4.21	4.30	4.02	4.02	4.35	4.67	8.95	6.20
Creative Entertainment Activities (ISIC Rev.4 Code: 90, 91, 92)	11.75	11.10	14.65	14.32	19.30	19.83	17.38	13.94	12.63	13.51	13.84
Sports & Recreation Activities (ISIC Rev.4 Code: 93)	9.15	9.34	7.31	8.31	11.73	11.72	11.88	11.33	15.24	13.52	13.78
Membership Organizations (ISIC Rev.4 Code: 94)	3.69	4.04	5.03	4.78	4.52	4.94	5.17	4.54	4.64	4.73	3.96
Repair Services (ISIC Rev.4 Code: 95)	13.91	15.18	15.59	13.09	10.01	15.74	26.87	25.85	23.29	26.17	26.14
Personal Service Activities (ISIC Rev.4 Code: 96)	53.87	52.63	52.06	50.21	46.89	46.41	47.81	48.01	45.80	45.86	45.78
Undifferentiated Services (ISIC Rev.4 Code: 97, 98)	22.46	22.38	22.63	23.06	22.73	22.80	22.69	22.73	22.44	22.00	21.83
All Services (Average)	26.88	26.97	27.23	27.74	27.13	26.94	26.52	26.18	25.81	25.53	25.21
Business Sector Services (Average)	19.45	19.45	20.02	20.99	20.48	20.27	20.35	20.58	20.16	19.40	19.54
Business Sector Services (Average excluding real estate)	17.66	17.63	18.06	18.39	18.10	17.98	18.14	18.17	17.89	17.94	17.23
All Industries (Average)	17.66	17.63	18.06	18.39	18.10	17.98	18.14	18.17	17.89	17.94	17.23

Note: Gross Profit Ratio is computed by dividing gross surplus by gross productive output in the corresponding service sector. Gross surplus and gross productive output data are secured from OECD databases.

Table 4.44 Gross profit ratio: Sweden

Industry	2000	2001	2002	2003	2004	2005	2006	2007	2008	2009	2010
Construction (ISIC Rev.4 Code: 41)	6.45	6.14	6.92	8.17	9.54	8.44	9.85	9.69	6.70	5.49	7.40
Automotive Retail & Repair (ISIC Rev.4 Code: 45)	17.22	16.42	17.69	19.04	23.84	22.70	22.74	21.93	19.16	18.65	23.00
Wholesale Trade (ISIC Rev.4 Code: 46)	17.60	16.94	18.58	18.78	20.37	20.51	21.26	21.30	21.70	21.64	22.87
Retail Trade (ISIC Rev.4 Code: 47)	7.96	7.37	9.54	9.77	11.66	12.39	13.78	13.68	11.76	13.32	14.45
Land Transport (ISIC Rev.4 Code: 49)	18.54	19.77	19.35	19.68	18.57	19.05	19.38	18.93	17.95	15.20	15.07
Water Transport (ISIC Rev.4 Code: 50)	11.35	11.75	8.15	9.20	11.92	12.79	12.29	9.87	10.28	-0.63	2.70
Air Transport (ISIC Rev.4 Code: 51)	3.29	0.68	4.46	5.30	5.03	7.84	8.41	7.80	7.45	9.48	8.57
Support Activities for Transport (ISIC Rev.4 Code: 52)											
Postal & Courier Activities (ISIC Rev.4 Code: 53)											
Hotel & Restaurants (ISIC Rev.4 Code: 55, 56)	9.75	8.17	8.02	6.51	6.27	7.45	7.76	7.38	5.63	4.16	2.93
Publishing Activities (ISIC Rev.4 Code: 58)	9.84	9.06	11.32	13.56	15.28	15.31	15.92	14.41	14.52	10.59	14.80
Audiovisual & Broadcasting (ISIC Rev.4 Code: 59, 60)	13.89	13.65	14.62	16.25	15.55	15.69	17.74	17.13	16.96	17.22	17.73
Telecommunications (ISIC Rev.4 Code: 61)	23.33	21.30	22.72	24.45	21.53	18.24	19.07	17.82	17.12	18.18	22.39

IT & Information Services (ISIC Rev.4 Code: 62, 63)	6.17	5.87	7.75	9.43	12.86	15.02	13.81	12.29	13.98	11.28	12.20
Financial Service Activities (ISIC Rev.4 Code: 64)	30.38	31.47	27.13	32.92	38.94	38.74	29.76	26.85	26.38	33.78	29.95
Insurance & Pensions (ISIC Rev.4 Code: 65)	33.53	32.74	32.44	35.01	36.72	39.91	34.17	34.76	35.47	36.64	33.39
Auxiliary Financial Services (ISIC Rev.4 Code: 66)	3.40	4.13	−14.51	−4.00	−5.51	−2.81	−8.46	1.64	−5.45	−3.14	−8.28
Legal, Accounting & Consultancy (ISIC Rev.4 Code: 69,70)	17.15	14.48	16.78	18.79	19.92	19.11	20.92	20.51	20.39	18.38	18.90
Architectural & Technical Services (ISIC Rev.4 Code: 71)											
Research & Development (ISIC Rev.4 Code: 72)											
Advertising & Market Research (ISIC Rev.4 Code: 73)	8.00	4.50	6.73	8.61	9.38	9.12	10.52	9.08	8.41	5.79	7.62
Other Professionals Services (ISIC Rev.4 Code: 74, 75)	15.73	13.99	15.22	15.30	15.37	15.33	16.49	15.35	15.87	15.60	15.98
Rental & Leasing Activities (ISIC Rev.4 Code: 77)	39.96	38.98	39.41	41.19	42.42	39.55	40.27	40.04	39.71	41.85	41.66
Employment Activities (ISIC Rev.4 Code: 78)	12.97	10.95	11.84	9.63	8.94	9.05	10.85	6.36	3.78	−0.53	1.21
Travel Service (ISIC Rev.4 Code: 79)	−0.16	−0.68	−1.20	2.21	2.33	2.36	3.01	0.72	0.13	2.05	3.95
Security & Building Services (ISIC Rev.4 Code: 80, 81, 82)	10.08	9.65	10.26	8.56	8.63	8.89	10.76	8.35	8.64	8.97	10.08
Public Administration & Defense (ISIC Rev.4 Code: 84)	8.73	9.38	9.80	9.81	9.70	10.02	10.11	10.49	10.88	11.28	11.22

(continued)

Table 4.44 (continued)

Industry	2000	2001	2002	2003	2004	2005	2006	2007	2008	2009	2010
Education (ISIC Rev.4 Code: 85)	8.32	5.91	8.08	7.72	7.45	7.82	7.92	6.77	7.55	8.04	8.24
Human Health Activities (ISIC Rev.4 Code: 86)	7.64	5.30	6.91	7.47	7.93	7.73	7.65	7.76	7.85	8.07	8.41
Residential Care & Social Work (ISIC Rev.4 Code: 87, 88)	2.61	2.20	2.95	2.24	1.53	2.07	2.26	0.49	1.28	1.65	1.61
Creative Entertainment Activities (ISIC Rev.4 Code: 90, 91, 92)	24.48	24.90	25.47	26.56	25.91	26.24	27.41	27.26	27.80	25.11	24.91
Sports & Recreation Activities (ISIC Rev.4 Code: 93)	8.55	8.40	9.23	10.56	10.15	10.42	11.99	11.64	11.26	11.13	12.57
Membership Organizations (ISIC Rev.4 Code: 94)	−2.07	−1.24	−1.54	−1.35	−1.53	−0.61	−0.34	−0.43	0.68	0.42	0.78
Repair Services (ISIC Rev.4 Code: 95)	9.59	6.83	10.64	12.76	14.36	15.92	16.64	15.90	16.34	16.00	15.75
Personal Service Activities (ISIC Rev.4 Code: 96)	35.52	35.82	37.24	37.00	37.98	38.61	38.24	37.61	37.54	37.37	37.01
Undifferentiated Services (ISIC Rev.4 Code: 97, 98)											
All Services (Average)	17.77	16.83	17.14	17.72	17.90	18.11	18.47	17.81	17.56	17.13	17.14
Business Sector Services (Average)	22.30	21.11	21.44	22.39	22.52	22.60	23.01	22.02	21.50	21.05	20.95
Business Sector Services (Average excluding real estate)	15.58	14.67	15.10	16.22	17.56	17.92	17.86	17.07	16.20	15.61	16.46
All Industries (Average)	16.22	15.14	15.63	16.18	16.52	16.42	16.85	16.34	15.40	15.02	16.34

Note: Gross Profit Ratio is computed by dividing gross surplus by gross productive output in the corresponding service sector. Gross surplus and gross productive output data are secured from OECD databases.

Table 4.45 Gross profit ratio: United States

Industry	2000	2001	2002	2003	2004	2005	2006	2007	2008	2009	2010
Construction (ISIC Rev.4 Code: 41)	16.52	16.41	16.57	16.52	17.08	17.22	16.39	15.50	13.22	14.44	15.14
Automotive Retail & Repair (ISIC Rev.4 Code: 45)											
Wholesale Trade (ISIC Rev.4 Code: 46)											
Retail Trade (ISIC Rev.4 Code: 47)											
Land Transport (ISIC Rev.4 Code: 49)	16.99	18.85	17.60	18.15	18.80	18.19	18.28	18.13	17.61	19.05	18.86
Water Transport (ISIC Rev.4 Code: 50)	14.33	14.40	12.07	13.11	13.00	12.00	19.61	19.19	18.57	21.54	20.82
Air Transport (ISIC Rev.4 Code: 51)	4.62	0.34	-1.64	6.78	6.75	6.81	8.11	6.43	6.10	8.29	8.82
Support Activities for Transport (ISIC Rev.4 Code: 52)	15.57	16.37	17.84	17.30	18.28	20.37	20.58	18.22	19.14	19.26	17.20
Postal & Courier Activities (ISIC Rev.4 Code: 53)	5.40	2.13	7.11	10.92	7.01	2.95	3.45	3.82	3.02	2.46	2.22
Hotel & Restaurants (ISIC Rev.4 Code: 55, 56)	14.09	14.75	15.28	14.50	14.34	14.66	14.41	13.70	12.65	12.27	13.67
Publishing Activities (ISIC Rev.4 Code: 58)	9.07	9.30	19.69	21.28	24.48	26.04	18.47	21.44	17.34	19.75	18.76
Audiovisual & Broadcasting (ISIC Rev.4 Code: 59, 60)											
Telecommunications (ISIC Rev.4 Code: 61)	23.71	24.86	25.61	25.10	27.38	29.95	29.23	31.17	32.73	31.53	32.32

(continued)

Table 4.45 (continued)

Industry	2000	2001	2002	2003	2004	2005	2006	2007	2008	2009	2010
IT & Information Services (ISIC Rev.4 Code: 62, 63)	−7.54	−0.52	12.49	14.87	18.81	15.78	18.03	15.71	15.68	17.54	15.88
Financial Service Activities (ISIC Rev.4 Code: 64)	34.78	35.33	38.76	37.48	30.94	30.75	29.08	26.43	28.90	31.61	37.05
Insurance & Pensions (ISIC Rev.4 Code: 65)	29.98	23.60	18.21	19.48	22.98	23.31	25.45	26.94	19.57	24.92	29.61
Auxiliary Financial Services (ISIC Rev.4 Code: 66)	0.55	11.06	10.33	7.27	4.70	8.18	7.97	0.16	−4.52	1.26	6.47
Legal, Accounting & Consultancy (ISIC Rev.4 Code: 69,70)											
Architectural & Technical Services (ISIC Rev.4 Code: 71)											
Research & Development (ISIC Rev.4 Code: 72)											
Advertising & Market Research (ISIC Rev.4 Code: 73)											
Other Professionals Services (ISIC Rev.4 Code: 74, 75)											
Rental & Leasing Activities (ISIC Rev.4 Code: 77)	49.94	56.75	56.27	49.48	44.38	43.42	51.18	48.79	51.11	51.84	57.48
Employment Activities (ISIC Rev.4 Code: 78)											
Travel Service (ISIC Rev.4 Code: 79)											
Security & Building Services (ISIC Rev.4 Code: 80, 81, 82)											

Public Administration & Defense (ISIC Rev.4 Code: 84)	13.08	12.42	11.69	11.09	10.92	11.05	11.21	11.09	10.94	11.10	10.96
Education (ISIC Rev.4 Code: 85)	6.12	5.59	5.71	5.76	6.00	6.01	6.08	6.17	6.31	6.77	6.79
Human Health Activities (ISIC Rev.4 Code: 86)											
Residential Care & Social Work (ISIC Rev.4 Code: 87, 88)											
Creative Entertainment Activities (ISIC Rev.4 Code: 90, 91, 92)											
Sports & Recreation Activities (ISIC Rev.4 Code: 93)											
Membership Organizations (ISIC Rev.4 Code: 94)											
Repair Services (ISIC Rev.4 Code: 95)											
Personal Service Activities (ISIC Rev.4 Code: 96)											
Undifferentiated Services (ISIC Rev.4 Code: 97, 98)											
All Services (Average)	21.30	21.91	22.40	22.20	21.89	21.91	21.74	21.58	21.44	22.38	22.56
Business Sector Services (Average)	24.86	26.38	27.29	27.26	26.74	26.68	26.37	26.20	26.31	27.91	27.83
Business Sector Services (Average excluding real estate)	17.50	19.01	20.14	20.11	19.80	19.93	19.82	19.26	18.86	20.29	21.63
All Industries (Average)	19.09	19.56	20.16	20.15	20.39	20.32	20.26	20.11	19.68	21.02	21.50

Note: Gross Profit Ratio is computed by dividing gross surplus by gross productive output in the corresponding service sector. Gross surplus and gross productive output data are secured from OECD databases.

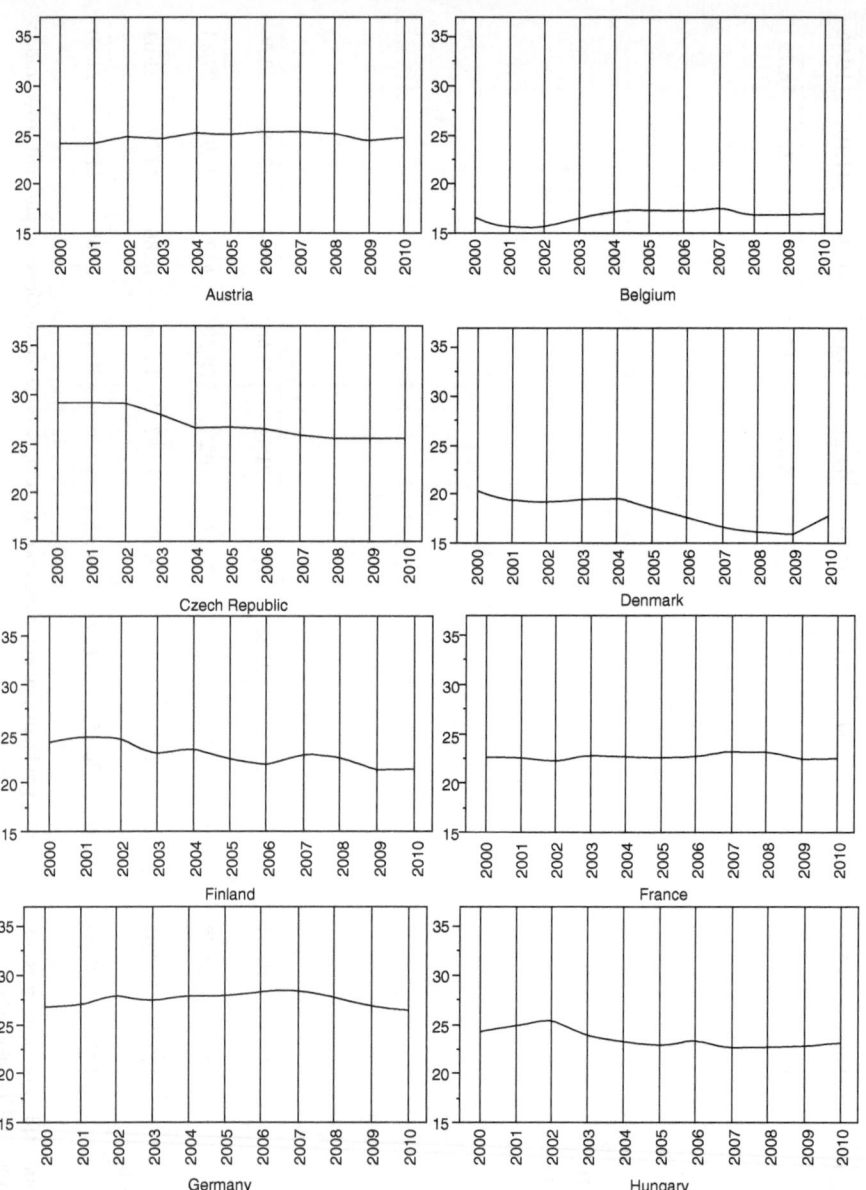

Fig. 4.3 Gross profit ratio for service sector across countries (2000–2010)

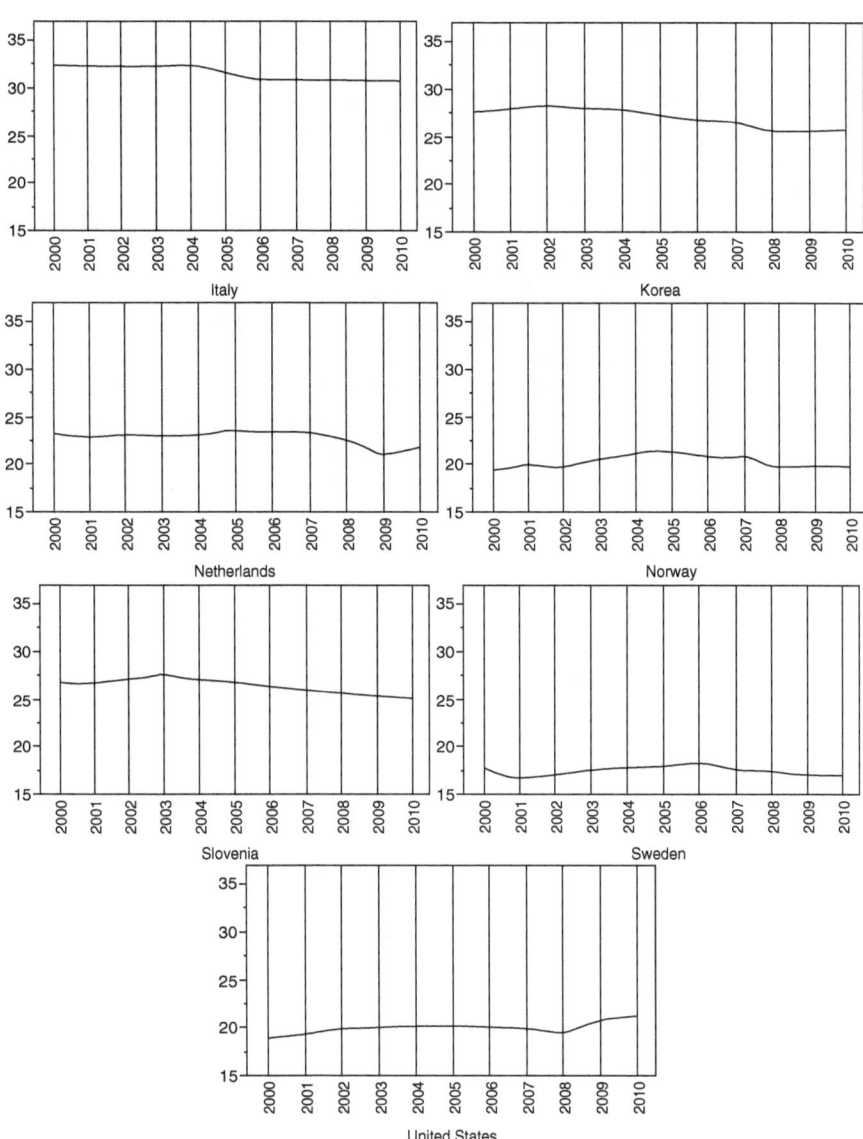

Fig. 4.3 (continued)

4.4   Domestic Global Market Ratio (DGM Ratio)

Domestic Global Market Ratio (DGM Ratio) compares the size of the domestic market relative to the global market in a particular industry and thus indicates the relative importance of a particular domestic market to the global market in terms of size. Firms located in markets with a large Domestic Global Market Ratio have an ability to grow and prosper in the domestic market compared to firms in markets with a smaller Domestic Global Market Ratio. As a result, the size of the Domestic Global Market Ratio can be a big motivator for a firm to seek foreign markets. Domestic Global Market Ratio was computed from OECD databases as follows: Initially, the gross productive output of each country by service sector was obtained, then the gross productive output of all the 15 OECD countries were added for each service sector. The DGM Ratio was then computed by dividing gross productive output of a country by the combined 15-country productive output in the corresponding service sector. In this computation, the OECD output is taken as a proxy for global market output. Therefore, it is likely that these numbers are inflated (i.e., likely to be lower when the total global market output is factored in). An additional caveat with the numbers presented is that currency conversion rates used across countries impact the values presented.

DGM Ratio numbers for the years 2000–2010 for 15 OECD countries across 11–12 industries are presented where data was available. Table 4.46 shows the DGM Ratio for Austria; Table 4.47 shows the DGM Ratio for Belgium; Table 4.48 shows the DGM Ratio for the Czech Republic; Table 4.49 shows the DGM Ratio for Denmark; Table 4.50 shows the DGM Ratio for Finland; Table 4.51 shows the DGM Ratio for France; Table 4.52 shows the DGM Ratio for Germany; Table 4.53 shows the DGM Ratio for Hungary; Table 4.54 shows the DGM Ratio for Italy; Table 4.55 shows the DGM Ratio for Korea; Table 4.56 shows the DGM Ratio for the Netherlands; Table 4.57 shows the DGM Ratio for Norway; Table 4.58 shows the DGM Ratio for Slovenia; Table 4.59 shows the DGM Ratio for Sweden; and Table 4.60 shows the DGM Ratio for the United States. Figure 4.4 graphically shows the average DGM Ratio for the services sector in each of the countries. It should be noted that in Figure 4.4, the y-axis coordinates for the United States varies compared to other country illustrations.

An overview of DGM Ratio during the years 2000–2010 across the fifteen OECD countries indicates that the United States DGM Ratio is the highest at 41.07, followed by Germany at 6.82. Slovenia had the lowest DGM Ratio (.08), with Hungary having the second lowest (.25). The United States, by virtue of being the largest economy in the world and its high per capita income, serves as the largest single market for many market segments in the service sector. Figure 4.5 presents a treemap of countries sized by the average of Domestic Global Market Ratio for the years 2000–2010. As one might expect, countries with a very small DGM Ratio are not visible in the figure. However, the relative dominance of the United States has had a 3.45 point reduction during the years 2000–2010 (from 42.85 to 39.4). This reduction is likely due to the higher relative growth of other countries, as the service sector continues to grow in the United States. Detailed statistics on the United States service sector is presented in Chapter 3.

Table 4.46 Domestic global market ratio: Austria

Industry	2000	2001	2002	2003	2004	2005	2006	2007	2008	2009	2010
Construction (ISIC Rev.4 Code: 41)	1.09	1.00	1.02	1.10	1.13	1.10	1.10	1.19	1.34	1.41	1.49
Automotive Retail & Repair (ISIC Rev.4 Code: 45)	2.67	2.53	2.37	2.37	2.46	2.38	2.33	2.35	2.36	2.47	2.73
Wholesale Trade (ISIC Rev.4 Code: 46)	2.47	2.45	2.50	2.42	2.34	2.43	2.51	2.47	2.39	2.49	2.47
Retail Trade (ISIC Rev.4 Code: 47)	2.29	2.27	2.30	2.25	2.27	2.31	2.31	2.33	2.37	2.46	2.58
Land Transport (ISIC Rev.4 Code: 49)	1.20	1.21	1.27	1.31	1.33	1.21	1.19	1.24	1.28	1.37	1.30
Water Transport (ISIC Rev.4 Code: 50)	0.06	0.06	0.07	0.06	0.05	0.05	0.06	0.05	0.06	0.06	0.06
Air Transport (ISIC Rev.4 Code: 51)	0.84	0.88	1.09	1.05	1.12	1.09	1.13	1.16	1.10	1.12	1.13
Support Activities for Transport (ISIC Rev.4 Code: 52)	0.56	0.55	0.56	0.60	0.81	1.09	1.15	1.15	1.19	1.12	1.18
Postal & Courier Activities (ISIC Rev.4 Code: 53)	1.08	0.96	0.94	1.00	1.02	1.15	1.16	1.19	1.21	1.21	1.27
Hotel & Restaurants (ISIC Rev.4 Code: 55, 56)	1.05	1.04	1.05	1.14	1.16	1.17	1.18	1.20	1.28	1.30	1.32
Publishing Activities (ISIC Rev.4 Code: 58)	0.32	0.36	0.36	0.40	0.40	0.40	0.42	0.42	0.47	0.46	0.47
Audiovisual & Broadcasting (ISIC Rev.4 Code: 59, 60)	1.34	1.27	1.29	1.27	1.36	1.36	1.32	1.32	1.31	1.32	1.37
Telecommunications (ISIC Rev.4 Code: 61)	0.35	0.58	0.54	0.61	0.66	0.64	0.63	0.63	0.62	0.58	0.56

(continued)

Table 4.46 (continued)

Industry	2000	2001	2002	2003	2004	2005	2006	2007	2008	2009	2010
IT & Information Services (ISIC Rev.4 Code: 62, 63)	0.72	0.73	0.82	0.86	0.91	0.90	0.88	0.86	0.84	0.85	0.84
Financial Service Activities (ISIC Rev.4 Code: 64)	0.88	0.81	0.73	0.75	0.76	0.76	0.83	0.88	0.89	0.74	0.76
Insurance & Pensions (ISIC Rev.4 Code: 65)	0.48	0.51	0.50	0.52	0.54	0.58	0.55	0.56	0.58	0.52	0.61
Auxiliary Financial Services (ISIC Rev.4 Code: 66)	0.12	0.16	0.21	0.26	0.27	0.30	0.30	0.31	0.31	0.27	0.28
Legal, Accounting & Consultancy (ISIC Rev.4 Code: 69,70)	1.23	1.22	1.30	1.38	1.44	1.55	1.59	1.68	1.73	1.88	1.96
Architectural & Technical Services (ISIC Rev.4 Code: 71)	2.31	2.34	2.33	2.37	2.33	2.48	2.62	2.51	2.44	2.37	2.44
Research & Development (ISIC Rev.4 Code: 72)	0.40	0.45	0.53	0.59	0.62	0.64	0.58	0.66	0.92	0.95	0.99
Advertising & Market Research (ISIC Rev.4 Code: 73)	2.22	2.37	2.51	2.48	2.54	2.59	2.71	2.77	2.75	2.80	2.91
Other Professionals Services (ISIC Rev.4 Code: 74, 75)	1.93	1.94	1.92	1.98	2.04	2.14	1.60	1.32	1.13	1.23	1.23
Rental & Leasing Activities (ISIC Rev.4 Code: 77)	0.58	0.56	0.59	0.67	0.72	0.74	0.81	0.85	0.95	0.96	0.96
Employment Activities (ISIC Rev.4 Code: 78)	1.39	1.56	1.63	1.69	1.86	2.05	2.05	1.98	2.01	2.02	2.11
Travel Service (ISIC Rev.4 Code: 79)	1.56	1.65	1.77	1.90	1.74	1.81	1.96	1.76	1.83	2.17	2.26
Security & Building Services (ISIC Rev.4 Code: 80, 81, 82)	1.26	1.29	1.31	1.35	1.36	1.37	1.47	1.50	1.56	1.63	1.72
Public Administration & Defense (ISIC Rev.4 Code: 84)	0.58	0.54	0.51	0.55	0.55	0.55	0.56	0.56	0.59	0.56	0.56

Education (ISIC Rev.4 Code: 85)	0.73	0.68	0.68	0.72	0.75	0.76	0.76	0.78	0.81	0.82	0.84
Human Health Activities (ISIC Rev.4 Code: 86)	2.03	2.10	2.03	2.03	2.02	2.02	2.00	2.01	2.02	1.98	1.94
Residential Care & Social Work (ISIC Rev.4 Code: 87, 88)	1.59	1.55	1.48	1.46	1.46	1.45	1.49	1.47	1.49	1.54	1.56
Creative Entertainment Activities (ISIC Rev.4 Code: 90, 91, 92)	1.80	1.70	1.67	1.66	1.76	1.88	1.82	1.83	1.91	2.02	2.03
Sports & Recreation Activities (ISIC Rev.4 Code: 93)	1.44	1.40	1.34	1.34	1.35	1.35	1.34	1.36	1.52	1.42	1.43
Membership Organizations (ISIC Rev.4 Code: 94)	2.59	2.58	2.44	2.42	2.39	2.41	2.38	2.33	2.39	2.41	2.39
Repair Services (ISIC Rev.4 Code: 95)	0.85	0.93	0.99	0.98	0.98	0.90	1.26	1.23	1.00	1.15	1.13
Personal Service Activities (ISIC Rev.4 Code: 96)	1.42	1.44	1.45	1.48	1.54	1.55	1.50	1.50	1.51	1.54	1.61
Undifferentiated Services (ISIC Rev.4 Code: 97, 98)	0.13	0.13	0.13	0.13	0.12	0.12	0.12	0.12	0.12	0.13	0.14
All Services (Average)	0.68	0.66	0.67	0.71	0.74	0.75	0.76	0.78	0.81	0.80	0.81
Business Sector Services (Average)	0.70	0.69	0.70	0.74	0.77	0.79	0.80	0.82	0.85	0.85	0.86
Business Sector Services (Average excluding real estate)	0.71	0.71	0.72	0.76	0.79	0.81	0.82	0.84	0.88	0.87	0.87
All Industries (Average)	0.56	0.57	0.59	0.62	0.63	0.62	0.62	0.63	0.63	0.63	0.59

Note: DGM (Domestic Global Market) Ratio is computed by dividing gross productive output of a country by the OECD productive output in the corresponding service sector. In this computation the OECD output is taken as a proxy for global market output. Therefore, it is likely that these numbers are inflated (i.e., likely to be lower when the total global market output is factored in). Gross productive output data is secured from OECD databases.

Table 4.47 Domestic global market ratio: Belgium

Industry	2000	2001	2002	2003	2004	2005	2006	2007	2008	2009	2010
Construction (ISIC Rev.4 Code: 41)	1.35	1.30	1.29	1.35	1.43	1.40	1.51	1.58	1.76	1.91	2.10
Automotive Retail & Repair (ISIC Rev.4 Code: 45)	3.81	4.00	4.10	3.92	3.77	3.84	3.80	3.85	3.68	3.45	3.84
Wholesale Trade (ISIC Rev.4 Code: 46)	4.08	3.94	4.00	3.95	3.99	4.11	3.96	3.88	3.76	3.62	3.91
Retail Trade (ISIC Rev.4 Code: 47)	3.17	3.18	3.27	3.11	3.11	3.07	3.05	3.09	3.19	3.29	3.47
Land Transport (ISIC Rev.4 Code: 49)	1.46	1.48	1.46	1.55	1.52	1.43	1.43	1.48	1.53	1.54	1.61
Water Transport (ISIC Rev.4 Code: 50)	1.61	1.56	1.45	1.54	1.93	2.31	2.40	2.09	2.29	2.18	2.17
Air Transport (ISIC Rev.4 Code: 51)	1.61	1.54	0.91	1.00	0.97	0.96	0.96	0.96	1.02	0.95	0.97
Support Activities for Transport (ISIC Rev.4 Code: 52)	2.85	2.81	2.67	2.82	2.95	3.44	3.43	3.56	3.68	3.50	3.66
Postal & Courier Activities (ISIC Rev.4 Code: 53)	1.36	1.26	1.26	1.43	1.54	1.56	1.63	1.71	1.78	1.73	1.83
Hotel & Restaurants (ISIC Rev.4 Code: 55, 56)	0.77	0.74	0.74	0.79	0.81	0.80	0.80	0.83	0.87	0.86	0.90
Publishing Activities (ISIC Rev.4 Code: 58)	0.50	0.48	0.45	0.50	0.53	0.46	0.46	0.47	0.49	0.46	0.48
Audiovisual & Broadcasting (ISIC Rev.4 Code: 59, 60)	2.04	2.02	1.85	1.92	1.97	2.00	1.94	1.91	2.02	2.05	2.23
Telecommunications (ISIC Rev.4 Code: 61)	0.41	0.74	0.75	0.84	0.92	0.90	0.91	0.89	0.92	0.90	0.89
IT & Information Services (ISIC Rev.4 Code: 62, 63)	0.93	1.01	0.99	1.02	1.06	1.00	1.06	1.10	1.12	1.16	1.15

Financial Service Activities (ISIC Rev.4 Code: 64)	1.30	1.15	1.11	1.07	1.21	1.21	1.25	1.27	1.23	1.21	1.35
Insurance & Pensions (ISIC Rev.4 Code: 65)	0.65	0.63	0.68	0.74	0.76	0.72	0.72	0.74	0.82	0.82	0.89
Auxiliary Financial Services (ISIC Rev.4 Code: 66)	0.58	0.59	0.69	0.76	0.78	0.78	0.72	0.76	0.99	0.90	0.83
Legal, Accounting & Consultancy (ISIC Rev.4 Code: 69,70)	4.55	4.79	4.70	4.69	5.04	5.33	5.39	5.38	5.53	5.83	5.99
Architectural & Technical Services (ISIC Rev.4 Code: 71)	2.87	2.84	2.83	2.71	2.69	2.56	2.62	2.55	2.61	2.73	2.92
Research & Development (ISIC Rev.4 Code: 72)	0.95	1.07	1.05	1.15	1.40	1.53	1.49	1.42	1.43	1.56	1.42
Advertising & Market Research (ISIC Rev.4 Code: 73)	4.65	4.92	4.69	4.85	4.81	4.93	4.93	4.81	4.92	4.75	4.91
Other Professionals Services (ISIC Rev.4 Code: 74, 75)	1.11	1.11	1.12	1.14	1.16	1.18	1.11	1.14	1.23	1.30	1.42
Rental & Leasing Activities (ISIC Rev.4 Code: 77)	0.65	0.73	0.74	0.74	0.80	0.79	0.77	0.86	0.90	0.91	0.97
Employment Activities (ISIC Rev.4 Code: 78)	4.07	3.77	3.96	3.94	4.06	4.00	3.82	3.49	3.33	3.56	3.77
Travel Service (ISIC Rev.4 Code: 79)	2.97	3.21	3.57	3.49	3.57	3.59	3.69	3.79	3.90	3.54	3.81
Security & Building Services (ISIC Rev.4 Code: 80, 81, 82)	2.97	2.92	3.04	3.20	3.05	3.20	3.07	3.29	3.23	3.34	3.52
Public Administration & Defense (ISIC Rev.4 Code: 84)	0.73	0.68	0.70	0.74	0.76	0.78	0.78	0.80	0.81	0.80	0.80
Education (ISIC Rev.4 Code: 85)	1.02	0.97	0.98	1.05	1.08	1.09	1.09	1.11	1.17	1.17	1.21

(continued)

Table 4.47 (continued)

Industry	2000	2001	2002	2003	2004	2005	2006	2007	2008	2009	2010
Human Health Activities (ISIC Rev.4 Code: 86)	2.89	2.85	2.77	2.81	2.84	2.81	2.77	2.78	2.86	2.91	2.93
Residential Care & Social Work (ISIC Rev.4 Code: 87, 88)	2.41	2.52	2.51	2.41	2.38	2.44	2.43	2.45	2.48	2.55	2.57
Creative Entertainment Activities (ISIC Rev.4 Code: 90, 91, 92)	1.86	1.79	1.70	1.55	1.82	1.87	1.87	1.88	1.85	1.60	1.65
Sports & Recreation Activities (ISIC Rev.4 Code: 93)	1.48	1.31	1.17	1.52	1.43	1.42	1.43	1.50	1.62	1.47	1.52
Membership Organizations (ISIC Rev.4 Code: 94)	2.37	2.29	2.38	2.35	2.47	2.54	2.64	2.64	2.87	3.64	4.04
Repair Services (ISIC Rev.4 Code: 95)	2.15	2.25	2.35	2.24	2.27	2.04	1.73	1.53	1.63	1.72	1.71
Personal Service Activities (ISIC Rev.4 Code: 96)	1.45	1.44	1.52	1.43	1.41	1.51	1.46	1.51	1.54	1.49	1.56
Undifferentiated Services (ISIC Rev.4 Code: 97, 98)	0.92	0.97	1.03	1.05	1.09	1.05	0.91	0.80	0.75	0.70	0.66
All Services (Average)	1.01	0.99	0.99	1.05	1.09	1.11	1.11	1.14	1.19	1.16	1.20
Business Sector Services (Average)	1.08	1.07	1.06	1.12	1.18	1.19	1.18	1.22	1.28	1.25	1.29
Business Sector Services (Average excluding real estate)	1.15	1.15	1.14	1.20	1.26	1.28	1.28	1.32	1.38	1.35	1.38
All Industries (Average)	0.87	0.85	0.73	0.73	0.78	0.70	0.66	0.67	0.71	0.79	0.81

Note: DGM (Domestic Global Market) Ratio is computed by dividing gross productive output of a country by the OECD productive output in the corresponding service sector. In this computation the OECD output is taken as a proxy for global market output. Therefore, it is likely that these numbers are inflated (i.e., likely to be lower when the total global market output is factored in). Gross productive output data is secured from OECD databases.

Table 4.48 Domestic global market ratio: Czech Republic

Industry	2000	2001	2002	2003	2004	2005	2006	2007	2008	2009	2010
Construction (ISIC Rev.4 Code: 41)	0.48	0.58	0.65	0.70	0.72	0.74	0.89	1.23	1.27	1.21	1.20
Automotive Retail & Repair (ISIC Rev.4 Code: 45)	0.64	0.78	0.87	0.95	1.01	1.08	1.26	1.59	1.76	1.58	1.50
Wholesale Trade (ISIC Rev.4 Code: 46)	0.92	1.02	1.01	1.06	1.17	1.18	1.33	1.51	1.59	1.38	1.55
Retail Trade (ISIC Rev.4 Code: 47)	0.68	0.82	0.99	0.99	1.07	1.13	1.26	1.49	1.49	1.49	1.54
Land Transport (ISIC Rev.4 Code: 49)	0.54	0.69	0.77	0.84	0.91	0.94	1.06	1.33	1.38	1.11	1.17
Water Transport (ISIC Rev.4 Code: 50)	0.04	0.03	0.02	0.02	0.02	0.02	0.02	0.02	0.03	0.02	0.02
Air Transport (ISIC Rev.4 Code: 51)	0.22	0.28	0.29	0.32	0.37	0.38	0.41	0.51	0.59	0.48	0.47
Support Activities for Transport (ISIC Rev.4 Code: 52)	0.57	0.68	0.69	0.81	0.82	0.85	0.94	1.23	1.38	1.26	1.31
Postal & Courier Activities (ISIC Rev.4 Code: 53)	0.20	0.26	0.30	0.33	0.37	0.40	0.47	0.57	0.54	0.62	0.67
Hotel & Restaurants (ISIC Rev.4 Code: 55, 56)	0.28	0.31	0.33	0.36	0.38	0.36	0.39	0.55	0.52	0.46	0.41
Publishing Activities (ISIC Rev.4 Code: 58)	0.15	0.19	0.20	0.19	0.20	0.25	0.28	0.32	0.31	0.33	0.29
Audiovisual & Broadcasting (ISIC Rev.4 Code: 59, 60)	0.50	0.58	0.70	0.77	0.83	0.89	1.07	1.31	1.15	1.06	1.14
Telecommunications (ISIC Rev.4 Code: 61)	0.24	0.29	0.33	0.37	0.38	0.41	0.46	0.55	0.51	0.48	0.45

(continued)

Table 4.48 (continued)

Industry	2000	2001	2002	2003	2004	2005	2006	2007	2008	2009	2010
IT & Information Services (ISIC Rev.4 Code: 62, 63)	0.16	0.22	0.26	0.29	0.33	0.40	0.47	0.65	0.67	0.66	0.67
Financial Service Activities (ISIC Rev.4 Code: 64)	0.18	0.20	0.22	0.25	0.26	0.31	0.34	0.43	0.39	0.39	0.43
Insurance & Pensions (ISIC Rev.4 Code: 65)	0.08	0.13	0.14	0.14	0.19	0.16	0.21	0.27	0.30	0.28	0.27
Auxiliary Financial Services (ISIC Rev.4 Code: 66)	0.06	0.10	0.11	0.11	0.11	0.09	0.09	0.17	0.16	0.15	0.16
Legal, Accounting & Consultancy (ISIC Rev.4 Code: 69,70)	0.38	0.44	0.54	0.60	0.52	0.53	0.55	0.72	0.85	0.85	0.85
Architectural & Technical Services (ISIC Rev.4 Code: 71)	1.00	1.30	1.30	1.44	1.55	1.61	1.65	2.22	2.43	2.42	2.35
Research & Development (ISIC Rev.4 Code: 72)	0.37	0.35	0.42	0.40	0.43	0.48	0.50	0.60	0.63	0.62	0.62
Advertising & Market Research (ISIC Rev.4 Code: 73)	1.15	1.28	1.34	1.71	1.85	2.09	2.00	2.46	2.88	2.53	2.46
Other Professionals Services (ISIC Rev.4 Code: 74, 75)	0.96	1.35	1.70	1.70	1.72	1.83	2.02	2.69	2.95	1.96	2.06
Rental & Leasing Activities (ISIC Rev.4 Code: 77)	0.05	0.06	0.08	0.09	0.10	0.12	0.14	0.19	0.23	0.21	0.22
Employment Activities (ISIC Rev.4 Code: 78)	0.07	0.10	0.09	0.11	0.12	0.13	0.17	0.40	0.53	0.32	0.34
Travel Service (ISIC Rev.4 Code: 79)	1.14	1.48	1.52	1.56	1.79	1.83	2.03	2.28	2.44	2.01	1.96
Security & Building Services (ISIC Rev.4 Code: 80, 81, 82)	0.50	0.64	0.70	0.75	0.75	0.76	0.79	1.07	1.08	1.14	0.98
Public Administration & Defense (ISIC Rev.4 Code: 84)	0.21	0.24	0.29	0.32	0.32	0.34	0.37	0.42	0.39	0.39	0.38

Education (ISIC Rev.4 Code: 85)	0.18	0.21	0.25	0.28	0.30	0.31	0.34	0.41	0.39	0.39	0.39
Human Health Activities (ISIC Rev.4 Code: 86)	0.49	0.56	0.61	0.64	0.67	0.71	0.76	0.91	0.89	0.90	0.90
Residential Care & Social Work (ISIC Rev.4 Code: 87, 88)	0.16	0.20	0.21	0.25	0.26	0.28	0.30	0.36	0.34	0.33	0.33
Creative Entertainment Activities (ISIC Rev.4 Code: 90, 91, 92)	0.97	1.11	1.00	1.13	1.20	1.24	1.44	1.66	1.68	1.65	1.60
Sports & Recreation Activities (ISIC Rev.4 Code: 93)	0.61	0.70	0.77	0.78	0.92	0.91	1.03	1.22	1.16	1.07	1.06
Membership Organizations (ISIC Rev.4 Code: 94)	0.45	0.46	0.54	0.49	0.62	0.70	0.74	0.92	0.77	0.71	0.74
Repair Services (ISIC Rev.4 Code: 95)	0.83	1.32	1.31	1.51	1.55	1.32	1.46	1.87	2.02	2.19	2.42
Personal Service Activities (ISIC Rev.4 Code: 96)	0.58	0.62	0.64	0.73	0.72	0.68	0.78	0.95	0.82	0.74	0.76
Undifferentiated Services (ISIC Rev.4 Code: 97, 98)	0.01	0.01	0.02	0.02	0.03	0.03	0.03	0.03	0.03	0.03	0.03
All Services (Average)	0.22	0.26	0.29	0.32	0.34	0.36	0.40	0.51	0.51	0.47	0.48
Business Sector Services (Average)	0.23	0.28	0.31	0.34	0.36	0.38	0.43	0.55	0.57	0.52	0.53
Business Sector Services (Average excluding real estate)	0.24	0.29	0.33	0.36	0.38	0.40	0.45	0.58	0.59	0.53	0.54
All Industries (Average)	0.23	0.26	0.31	0.35	0.37	0.38	0.42	0.45	0.51	0.46	0.42

Note: DGM (Domestic Global Market) Ratio is computed by dividing gross productive output of a country by the OECD productive output in the corresponding service sector. In this computation the OECD output is taken as a proxy for global market output. Therefore, it is likely that these numbers are inflated (i.e., likely to be lower when the total global market output is factored in). Gross productive output data is secured from OECD databases.

Table 4.49 Domestic global market ratio: Denmark

Industry	2000	2001	2002	2003	2004	2005	2006	2007	2008	2009	2010
Construction (ISIC Rev.4 Code: 41)	1.06	0.98	0.94	1.00	0.96	0.85	0.85	0.90	1.00	1.08	1.10
Automotive Retail & Repair (ISIC Rev.4 Code: 45)	2.39	2.52	2.70	2.91	2.81	2.74	2.70	2.56	2.57	2.67	2.47
Wholesale Trade (ISIC Rev.4 Code: 46)	2.09	2.10	2.03	2.15	2.22	2.14	2.12	2.21	2.28	2.35	2.34
Retail Trade (ISIC Rev.4 Code: 47)	2.84	2.85	2.82	2.93	2.92	2.83	2.78	2.79	2.79	2.77	2.70
Land Transport (ISIC Rev.4 Code: 49)	0.97	0.98	0.96	1.05	1.01	0.98	0.98	1.02	1.05	1.05	0.95
Water Transport (ISIC Rev.4 Code: 50)	1.09	1.17	1.17	1.28	1.49	1.53	1.60	1.84	2.04	1.91	1.88
Air Transport (ISIC Rev.4 Code: 51)	0.86	0.88	1.02	1.08	1.22	1.24	1.25	1.32	1.35	1.35	1.30
Support Activities for Transport (ISIC Rev.4 Code: 52)	1.65	1.69	1.77	1.94	2.03	1.98	1.99	2.12	2.21	2.14	2.06
Postal & Courier Activities (ISIC Rev.4 Code: 53)	1.22	1.20	1.31	1.47	1.56	1.54	1.48	1.53	1.66	1.62	1.50
Hotel & Restaurants (ISIC Rev.4 Code: 55, 56)	0.66	0.65	0.64	0.69	0.70	0.67	0.65	0.68	0.72	0.71	0.69
Publishing Activities (ISIC Rev.4 Code: 58)	0.77	0.73	0.71	0.80	0.83	0.79	0.78	0.87	1.00	1.08	0.98
Audiovisual & Broadcasting (ISIC Rev.4 Code: 59, 60)	2.89	2.85	2.67	2.73	2.71	2.64	2.56	2.53	2.61	2.73	2.51
Telecommunications (ISIC Rev.4 Code: 61)	0.34	0.62	0.68	0.80	0.84	0.79	0.76	0.80	0.84	0.89	0.79
IT & Information Services (ISIC Rev.4 Code: 62, 63)	0.76	0.83	0.84	0.94	0.95	0.96	0.91	0.97	1.02	1.07	0.99

Financial Service Activities (ISIC Rev.4 Code: 64)	0.98	0.95	0.97	1.11	1.17	1.11	1.05	1.05	1.07	1.04	1.05
Insurance & Pensions (ISIC Rev.4 Code: 65)	0.86	0.82	0.86	1.03	1.08	1.01	0.98	0.98	1.01	1.07	1.01
Auxiliary Financial Services (ISIC Rev.4 Code: 66)	0.39	0.42	0.47	0.56	0.61	0.56	0.49	0.51	0.59	0.54	0.48
Legal, Accounting & Consultancy (ISIC Rev.4 Code: 69,70)	2.85	2.94	3.00	3.25	3.18	3.08	3.07	3.19	3.24	2.92	2.69
Architectural & Technical Services (ISIC Rev.4 Code: 71)	2.91	2.95	2.91	2.93	2.76	2.68	2.55	2.62	2.67	2.77	2.91
Research & Development (ISIC Rev.4 Code: 72)	1.85	1.74	1.83	1.89	1.97	1.76	1.72	1.85	2.06	1.90	1.74
Advertising & Market Research (ISIC Rev.4 Code: 73)	2.33	2.30	2.24	2.33	2.36	2.23	2.24	2.32	2.35	2.39	2.25
Other Professionals Services (ISIC Rev.4 Code: 74, 75)	2.64	2.63	2.78	2.69	2.54	2.51	2.51	2.55	2.41	2.45	2.39
Rental & Leasing Activities (ISIC Rev.4 Code: 77)	1.39	1.43	1.44	1.63	1.71	1.63	1.54	1.65	1.61	1.51	1.43
Employment Activities (ISIC Rev.4 Code: 78)	1.44	1.49	1.42	1.54	1.65	1.62	1.76	1.93	2.05	2.15	2.35
Travel Service (ISIC Rev.4 Code: 79)	3.57	3.60	3.24	3.35	3.48	3.39	3.21	3.22	3.39	3.60	3.50
Security & Building Services (ISIC Rev.4 Code: 80, 81, 82)	2.44	2.36	2.40	2.51	2.52	2.56	2.41	2.46	2.55	2.59	2.51
Public Administration & Defense (ISIC Rev.4 Code: 84)	0.75	0.72	0.72	0.80	0.82	0.78	0.75	0.78	0.82	0.81	0.75
Education (ISIC Rev.4 Code: 85)	0.84	0.81	0.85	0.92	0.98	0.94	0.89	0.93	0.96	0.96	0.90

(continued)

Table 4.49 (continued)

Industry	2000	2001	2002	2003	2004	2005	2006	2007	2008	2009	2010
Human Health Activities (ISIC Rev.4 Code: 86)	2.79	2.79	2.83	2.94	2.89	2.72	2.66	2.69	2.76	2.80	2.57
Residential Care & Social Work (ISIC Rev.4 Code: 87, 88)	2.29	2.26	2.27	2.37	2.36	2.28	2.20	2.24	2.27	2.33	2.16
Creative Entertainment Activities (ISIC Rev.4 Code: 90, 91, 92)	2.96	2.98	2.89	2.94	2.92	2.76	2.66	2.72	2.77	2.77	2.62
Sports & Recreation Activities (ISIC Rev.4 Code: 93)	2.37	2.38	2.33	2.44	2.49	2.39	2.32	2.32	2.51	2.53	2.36
Membership Organizations (ISIC Rev.4 Code: 94)	3.45	3.53	3.56	3.82	3.80	3.53	3.42	3.44	3.54	3.57	3.41
Repair Services (ISIC Rev.4 Code: 95)	1.04	1.01	0.92	0.96	0.96	1.00	0.97	0.98	1.12	0.98	0.93
Personal Service Activities (ISIC Rev.4 Code: 96)	4.26	4.30	4.30	4.57	4.67	4.50	4.40	4.53	4.64	4.72	4.35
Undifferentiated Services (ISIC Rev.4 Code: 97, 98)	1.39	1.41	1.43	1.56	1.57	1.51	1.46	1.50	1.48	1.49	1.45
All Services (Average)	0.89	0.88	0.90	1.00	1.03	0.99	0.96	1.02	1.07	1.06	0.99
Business Sector Services (Average)	0.90	0.90	0.91	1.02	1.05	1.01	0.98	1.04	1.10	1.09	1.02
Business Sector Services (Average excluding real estate)	0.88	0.89	0.90	1.01	1.04	1.01	0.98	1.03	1.09	1.08	1.00
All Industries (Average)	0.44	0.46	0.49	0.54	0.55	0.54	0.53	0.55	0.56	0.54	0.47

Note: DGM (Domestic Global Market) Ratio is computed by dividing gross productive output of a country by the OECD productive output in the corresponding service sector. In this computation the OECD output is taken as a proxy for global market output. Therefore, it is likely that these numbers are inflated (i.e., likely to be lower when the total global market output is factored in). Gross productive output data is secured from OECD databases.

Table 4.50 Domestic global market ratio: Finland

Industry	2000	2001	2002	2003	2004	2005	2006	2007	2008	2009	2010
Construction (ISIC Rev.4 Code: 41)	0.70	0.67	0.67	0.70	0.74	0.76	0.76	0.87	0.95	0.89	0.96
Automotive Retail & Repair (ISIC Rev.4 Code: 45)	1.27	1.26	1.29	1.37	1.46	1.51	1.52	1.52	1.59	1.52	1.67
Wholesale Trade (ISIC Rev.4 Code: 46)	1.33	1.31	1.32	1.30	1.30	1.33	1.34	1.33	1.22	1.20	1.14
Retail Trade (ISIC Rev.4 Code: 47)	1.05	1.09	1.13	1.18	1.21	1.26	1.32	1.30	1.35	1.36	1.39
Land Transport (ISIC Rev.4 Code: 49)	0.70	0.69	0.70	0.75	0.74	0.75	0.75	0.77	0.85	0.89	0.87
Water Transport (ISIC Rev.4 Code: 50)	1.41	1.33	1.37	1.44	1.38	1.29	1.24	1.27	1.21	1.24	1.17
Air Transport (ISIC Rev.4 Code: 51)	0.49	0.55	0.59	0.57	0.66	0.70	0.71	0.77	0.79	0.75	0.80
Support Activities for Transport (ISIC Rev.4 Code: 52)	0.87	0.83	0.81	0.88	0.91	0.93	1.01	1.05	0.94	0.84	0.92
Postal & Courier Activities (ISIC Rev.4 Code: 53)	0.62	0.59	0.62	0.66	0.69	0.71	0.71	0.72	0.76	0.74	0.77
Hotel & Restaurants (ISIC Rev.4 Code: 55, 56)	0.34	0.34	0.34	0.36	0.38	0.39	0.38	0.38	0.42	0.41	0.41
Publishing Activities (ISIC Rev.4 Code: 58)	0.39	0.38	0.41	0.49	0.49	0.49	0.48	0.51	0.56	0.58	0.59
Audiovisual & Broadcasting (ISIC Rev.4 Code: 59, 60)	0.94	0.94	0.92	0.93	1.01	0.98	0.96	1.01	1.00	0.97	0.98
Telecommunications (ISIC Rev.4 Code: 61)	0.29	0.51	0.55	0.61	0.65	0.57	0.54	0.40	0.40	0.38	0.39

(continued)

Table 4.50 (continued)

Industry	2000	2001	2002	2003	2004	2005	2006	2007	2008	2009	2010
IT & Information Services (ISIC Rev.4 Code: 62, 63)	0.46	0.53	0.55	0.62	0.70	0.71	0.76	0.74	0.78	0.73	0.72
Financial Service Activities (ISIC Rev.4 Code: 64)	0.27	0.25	0.23	0.23	0.23	0.23	0.27	0.30	0.29	0.28	0.27
Insurance & Pensions (ISIC Rev.4 Code: 65)	0.31	0.20	0.27	0.14	0.16	0.12	0.15	0.19	0.19	0.17	0.20
Auxiliary Financial Services (ISIC Rev.4 Code: 66)	0.10	0.11	0.10	0.12	0.12	0.12	0.11	0.12	0.10	0.09	0.10
Legal, Accounting & Consultancy (ISIC Rev.4 Code: 69,70)	0.52	0.48	0.48	0.43	0.45	0.46	0.47	0.55	0.56	0.58	0.65
Architectural & Technical Services (ISIC Rev.4 Code: 71)	1.40	1.41	1.40	1.54	1.66	1.69	1.77	1.81	1.82	1.62	1.71
Research & Development (ISIC Rev.4 Code: 72)	0.88	0.92	0.99	1.09	1.03	1.06	1.03	0.98	1.06	1.10	1.16
Advertising & Market Research (ISIC Rev.4 Code: 73)	0.98	0.91	0.92	0.90	0.88	0.90	0.90	0.84	0.73	0.75	0.75
Other Professionals Services (ISIC Rev.4 Code: 74, 75)	0.85	0.90	0.87	0.95	1.04	1.00	0.97	1.04	1.00	1.17	1.16
Rental & Leasing Activities (ISIC Rev.4 Code: 77)	0.11	0.12	0.14	0.15	0.17	0.17	0.19	0.24	0.22	0.24	0.25
Employment Activities (ISIC Rev.4 Code: 78)	0.50	0.50	0.57	0.63	0.72	0.88	0.92	0.97	1.10	1.20	1.24
Travel Service (ISIC Rev.4 Code: 79)	0.50	0.48	0.58	0.59	0.59	0.58	0.58	0.62	0.65	0.72	0.70
Security & Building Services (ISIC Rev.4 Code: 80, 81, 82)	0.85	0.86	0.93	0.96	1.00	0.99	1.04	1.11	1.31	1.32	1.37
Public Administration & Defense (ISIC Rev.4 Code: 84)	0.39	0.37	0.38	0.41	0.43	0.43	0.43	0.45	0.47	0.45	0.45

Education (ISIC Rev.4 Code: 85)	0.47	0.45	0.46	0.49	0.52	0.52	0.51	0.52	0.54	0.55	0.57
Human Health Activities (ISIC Rev.4 Code: 86)	1.26	1.27	1.26	1.28	1.32	1.35	1.36	1.37	1.44	1.47	1.48
Residential Care & Social Work (ISIC Rev.4 Code: 87, 88)	1.91	1.95	1.92	1.92	1.94	1.96	1.97	1.98	2.05	2.12	2.12
Creative Entertainment Activities (ISIC Rev.4 Code: 90, 91, 92)	1.17	1.14	1.12	1.13	1.13	1.15	1.13	1.18	1.24	1.24	1.29
Sports & Recreation Activities (ISIC Rev.4 Code: 93)	1.28	1.36	1.34	1.44	1.42	1.46	1.43	1.42	1.43	1.47	1.47
Membership Organizations (ISIC Rev.4 Code: 94)	1.82	1.87	1.87	1.90	1.96	1.99	2.05	2.09	2.14	2.11	2.19
Repair Services (ISIC Rev.4 Code: 95)	1.00	1.08	1.06	1.24	1.23	1.24	1.38	1.24	1.31	1.28	1.22
Personal Service Activities (ISIC Rev.4 Code: 96)	0.56	0.58	0.61	0.63	0.65	0.67	0.68	0.72	0.77	0.77	0.79
Undifferentiated Services (ISIC Rev.4 Code: 97, 98)	0.10	0.09	0.10	0.13	0.18	0.20	0.19	0.19	0.18	0.23	0.26
All Services (Average)	0.41	0.41	0.41	0.44	0.47	0.47	0.47	0.49	0.51	0.50	0.51
Business Sector Services (Average)	0.40	0.40	0.41	0.44	0.46	0.45	0.46	0.48	0.50	0.49	0.49
Business Sector Services (Average excluding real estate)	0.39	0.39	0.40	0.42	0.44	0.44	0.45	0.47	0.48	0.47	0.47
All Industries (Average)	0.40	0.40	0.41	0.42	0.43	0.42	0.42	0.43	0.42	0.40	0.38

Note: DGM (Domestic Global Market) Ratio is computed by dividing gross productive output of a country by the OECD productive output in the corresponding service sector. In this computation the OECD output is taken as a proxy for global market output. Therefore, it is likely that these numbers are inflated (i.e., likely to be lower when the total global market output is factored in). Gross productive output data is secured from OECD databases.

Table 4.51 Domestic global market ratio: France

Industry	2000	2001	2002	2003	2004	2005	2006	2007	2008	2009	2010
Construction (ISIC Rev.4 Code: 41)	6.15	6.05	6.22	6.58	6.93	6.99	7.25	7.84	8.39	8.50	8.67
Automotive Retail & Repair (ISIC Rev.4 Code: 45)	12.55	12.36	11.91	12.05	12.23	12.31	12.28	12.52	12.38	12.35	13.00
Wholesale Trade (ISIC Rev.4 Code: 46)	15.14	15.73	15.87	16.15	16.28	16.28	15.93	15.49	15.39	15.56	15.91
Retail Trade (ISIC Rev.4 Code: 47)	14.69	14.87	15.09	15.46	15.63	15.59	15.51	15.27	15.45	15.98	16.05
Land Transport (ISIC Rev.4 Code: 49)	6.56	6.49	6.58	6.90	7.15	7.08	7.05	7.34	7.30	7.60	7.49
Water Transport (ISIC Rev.4 Code: 50)	4.55	4.40	4.71	4.69	5.26	5.39	5.75	5.94	5.82	5.81	6.75
Air Transport (ISIC Rev.4 Code: 51)	4.92	5.14	5.63	5.21	5.32	5.59	5.65	5.86	6.08	6.02	6.06
Support Activities for Transport (ISIC Rev.4 Code: 52)	7.72	7.45	7.45	7.84	7.94	7.74	7.60	7.64	7.60	7.95	8.14
Postal & Courier Activities (ISIC Rev.4 Code: 53)	5.64	5.39	5.45	5.84	6.23	6.31	6.33	6.45	6.51	6.34	6.60
Hotel & Restaurants (ISIC Rev.4 Code: 55, 56)	4.83	4.71	4.77	5.13	5.24	5.23	5.20	5.35	5.47	5.48	5.55
Publishing Activities (ISIC Rev.4 Code: 58)	3.66	3.63	3.61	3.90	4.08	4.05	4.10	4.14	4.41	4.24	4.43
Audiovisual & Broadcasting (ISIC Rev.4 Code: 59, 60)	14.86	15.33	15.70	15.93	16.07	15.84	15.47	15.18	15.58	15.51	15.84
Telecommunications (ISIC Rev.4 Code: 61)	2.12	3.87	4.04	4.44	4.64	4.81	4.73	4.90	5.18	5.21	5.36
IT & Information Services (ISIC Rev.4 Code: 62, 63)	6.71	6.85	6.85	7.29	7.58	7.54	7.71	7.51	7.65	7.27	7.29

Financial Service Activities (ISIC Rev.4 Code: 64)	6.31	5.45	5.32	5.18	5.28	5.07	4.97	5.31	5.04	5.32	5.60
Insurance & Pensions (ISIC Rev.4 Code: 65)	3.36	2.86	3.12	3.60	3.73	3.78	4.01	3.99	4.12	4.19	4.42
Auxiliary Financial Services (ISIC Rev.4 Code: 66)	2.01	2.24	2.57	2.87	3.02	2.92	3.09	3.38	3.71	3.46	3.36
Legal, Accounting & Consultancy (ISIC Rev.4 Code: 69,70)	15.85	16.49	17.00	16.75	17.12	17.09	17.41	17.12	17.03	17.61	18.15
Architectural & Technical Services (ISIC Rev.4 Code: 71)	17.19	16.79	17.29	18.10	18.64	19.07	19.54	18.89	18.82	18.25	19.56
Research & Development (ISIC Rev.4 Code: 72)	32.00	32.49	31.85	31.22	30.59	30.68	30.68	30.24	29.10	30.39	30.30
Advertising & Market Research (ISIC Rev.4 Code: 73)	12.35	12.24	12.04	11.91	12.25	12.27	12.34	11.98	11.89	11.85	12.22
Other Professionals Services (ISIC Rev.4 Code: 74, 75)	8.14	8.34	8.48	8.79	9.00	8.98	9.31	9.25	9.43	9.05	9.08
Rental & Leasing Activities (ISIC Rev.4 Code: 77)	8.55	8.55	9.16	9.13	9.36	9.26	9.52	9.72	9.87	10.00	10.74
Employment Activities (ISIC Rev.4 Code: 78)	29.68	28.23	27.83	27.54	26.62	25.60	24.01	22.31	21.32	20.78	21.45
Travel Service (ISIC Rev.4 Code: 79)	6.39	6.65	7.64	7.77	7.98	7.83	7.88	7.78	7.99	7.75	8.05
Security & Building Services (ISIC Rev.4 Code: 80, 81, 82)	20.17	20.75	20.80	20.43	20.66	20.33	21.24	20.93	20.64	19.92	20.29
Public Administration & Defense (ISIC Rev.4 Code: 84)	4.96	4.55	4.43	4.67	4.89	4.86	4.80	4.88	4.83	4.75	4.81
Education (ISIC Rev.4 Code: 85)	5.45	5.10	5.21	5.52	5.64	5.63	5.57	5.63	5.77	5.69	5.85

(continued)

Table 4.51 (continued)

Industry	2000	2001	2002	2003	2004	2005	2006	2007	2008	2009	2010
Human Health Activities (ISIC Rev.4 Code: 86)	14.59	14.31	14.23	14.42	14.57	14.49	14.35	14.28	14.09	14.17	14.07
Residential Care & Social Work (ISIC Rev.4 Code: 87, 88)	16.15	16.23	15.86	15.53	15.89	16.10	16.38	16.17	15.98	16.05	16.07
Creative Entertainment Activities (ISIC Rev.4 Code: 90, 91, 92)	13.16	13.35	13.90	14.41	14.39	14.64	15.00	14.78	14.80	15.28	16.00
Sports & Recreation Activities (ISIC Rev.4 Code: 93)	15.66	15.88	15.44	15.58	15.61	15.55	15.52	15.21	15.36	16.07	16.33
Membership Organizations (ISIC Rev.4 Code: 94)	11.47	11.65	11.36	11.01	10.98	11.03	11.19	11.13	11.07	11.27	11.49
Repair Services (ISIC Rev.4 Code: 95)	27.40	27.87	28.59	28.58	28.56	29.37	29.57	29.26	28.98	29.35	29.41
Personal Service Activities (ISIC Rev.4 Code: 96)	8.68	9.17	9.36	9.40	9.52	9.37	9.41	9.36	9.42	9.26	9.38
Undifferentiated Services (ISIC Rev.4 Code: 97, 98)	8.36	8.43	8.84	9.24	9.69	9.99	9.80	9.52	9.66	9.79	10.34
All Services (Average)	5.29	5.15	5.21	5.54	5.76	5.74	5.77	5.90	6.04	5.98	6.06
Business Sector Services (Average)	5.36	5.28	5.38	5.72	5.93	5.90	5.94	6.09	6.28	6.24	6.31
Business Sector Services (Average excluding real estate)	5.41	5.39	5.50	5.83	6.03	5.99	6.01	6.14	6.32	6.30	6.32
All Industries (Average)	4.10	4.14	4.25	4.41	4.43	4.28	4.19	4.15	4.04	4.05	3.79

Note: DGM (Domestic Global Market) Ratio is computed by dividing gross productive output of a country by the OECD productive output in the corresponding service sector. In this computation the OECD output is taken as a proxy for global market output. Therefore, it is likely that these numbers are inflated (i.e., likely to be lower when the total global market output is factored in). Gross productive output data is secured from OECD databases.

Table 4.52 Domestic global market ratio: Germany

Industry	2000	2001	2002	2003	2004	2005	2006	2007	2008	2009	2010
Construction (ISIC Rev.4 Code: 41)	8.24	7.37	6.84	6.80	6.45	5.83	5.86	6.02	6.51	7.06	7.80
Automotive Retail & Repair (ISIC Rev.4 Code: 45)	18.57	18.97	19.55	19.79	18.80	18.81	18.67	17.12	16.69	17.46	17.46
Wholesale Trade (ISIC Rev.4 Code: 46)	16.26	15.82	14.74	14.65	14.86	14.71	14.65	14.75	14.77	15.33	16.52
Retail Trade (ISIC Rev.4 Code: 47)	22.09	21.47	20.40	19.92	19.53	19.41	19.22	18.63	18.13	18.09	19.10
Land Transport (ISIC Rev.4 Code: 49)	7.53	7.35	6.93	7.14	6.78	6.72	6.77	6.79	6.83	6.85	6.73
Water Transport (ISIC Rev.4 Code: 50)	8.51	8.81	8.47	8.72	9.97	10.47	11.10	12.28	13.26	12.47	13.28
Air Transport (ISIC Rev.4 Code: 51)	6.70	6.63	7.37	7.35	8.16	8.49	8.67	8.83	8.75	8.80	9.19
Support Activities for Transport (ISIC Rev.4 Code: 52)	12.83	12.70	12.85	13.18	13.58	13.59	13.81	14.20	14.33	13.96	14.55
Postal & Courier Activities (ISIC Rev.4 Code: 53)	9.51	9.06	9.53	10.00	10.45	10.59	10.22	10.24	10.76	10.60	10.62
Hotel & Restaurants (ISIC Rev.4 Code: 55, 56)	5.15	4.87	4.61	4.68	4.69	4.59	4.50	4.56	4.67	4.67	4.89
Publishing Activities (ISIC Rev.4 Code: 58)	5.95	5.46	5.13	5.42	5.55	5.44	5.41	5.84	6.50	7.07	6.91
Audiovisual & Broadcasting (ISIC Rev.4 Code: 59, 60)	22.42	21.44	19.37	18.59	18.17	18.09	17.72	16.89	16.96	17.82	17.72
Telecommunications (ISIC Rev.4 Code: 61)	2.65	4.64	4.90	5.43	5.64	5.42	5.27	5.35	5.47	5.84	5.61

(continued)

Table 4.52 (continued)

Industry	2000	2001	2002	2003	2004	2005	2006	2007	2008	2009	2010
IT & Information Services (ISIC Rev.4 Code: 62, 63)	5.90	6.27	6.08	6.38	6.39	6.57	6.31	6.49	6.63	7.02	7.00
Financial Service Activities (ISIC Rev.4 Code: 64)	7.60	7.18	7.01	7.55	7.85	7.60	7.30	7.00	6.91	6.82	7.44
Insurance & Pensions (ISIC Rev.4 Code: 65)	6.68	6.14	6.22	6.99	7.23	6.90	6.75	6.54	6.56	7.01	7.15
Auxiliary Financial Services (ISIC Rev.4 Code: 66)	2.99	3.14	3.40	3.79	4.10	3.86	3.43	3.39	3.81	3.52	3.37
Legal, Accounting & Consultancy (ISIC Rev.4 Code: 69,70)	22.13	22.10	21.75	22.12	21.32	21.17	21.29	21.33	21.00	19.07	19.05
Architectural & Technical Services (ISIC Rev.4 Code: 71)	22.62	22.20	21.12	19.93	18.48	18.37	17.65	17.53	17.30	18.07	20.56
Research & Development (ISIC Rev.4 Code: 72)	14.38	13.06	13.29	12.88	13.17	12.06	11.91	12.34	13.40	12.41	12.32
Advertising & Market Research (ISIC Rev.4 Code: 73)	18.10	17.34	16.22	15.86	15.80	15.27	15.53	15.51	15.24	15.60	15.94
Other Professionals Services (ISIC Rev.4 Code: 74, 75)	20.52	19.80	20.11	18.30	16.98	17.22	17.41	17.03	15.61	16.01	16.93
Rental & Leasing Activities (ISIC Rev.4 Code: 77)	10.81	10.73	10.46	11.10	11.47	11.20	10.64	11.01	10.44	9.84	10.10
Employment Activities (ISIC Rev.4 Code: 78)	11.19	11.18	10.26	10.51	11.05	11.09	12.19	12.93	13.33	14.04	16.61
Travel Service (ISIC Rev.4 Code: 79)	27.77	27.07	23.50	22.78	23.31	23.29	22.26	21.52	21.98	23.53	24.77
Security & Building Services (ISIC Rev.4 Code: 80, 81, 82)	18.96	17.77	17.42	17.11	16.86	17.54	16.70	16.46	16.53	16.89	17.73
Public Administration & Defense (ISIC Rev.4 Code: 84)	5.86	5.41	5.24	5.41	5.46	5.34	5.21	5.22	5.33	5.30	5.33

Education (ISIC Rev.4 Code: 85)	6.53	6.12	6.14	6.28	6.55	6.45	6.14	6.23	6.26	6.26	6.37
Human Health Activities (ISIC Rev.4 Code: 86)	21.69	20.98	20.49	20.00	19.39	18.67	18.43	17.96	17.93	18.32	18.18
Residential Care & Social Work (ISIC Rev.4 Code: 87, 88)	17.76	17.03	16.45	16.12	15.77	15.63	15.22	14.97	14.70	15.19	15.24
Creative Entertainment Activities (ISIC Rev.4 Code: 90, 91, 92)	23.03	22.40	20.91	19.98	19.55	18.94	18.45	18.16	17.97	18.07	18.52
Sports & Recreation Activities (ISIC Rev.4 Code: 93)	18.44	17.91	16.88	16.58	16.65	16.41	16.08	15.52	16.31	16.55	16.70
Membership Organizations (ISIC Rev.4 Code: 94)	26.77	26.54	25.80	25.99	25.47	24.24	23.67	22.98	22.99	23.32	24.13
Repair Services (ISIC Rev.4 Code: 95)	8.06	7.60	6.67	6.52	6.44	6.84	6.74	6.54	7.26	6.41	6.56
Personal Service Activities (ISIC Rev.4 Code: 96)	33.10	32.36	31.15	31.13	31.30	30.90	30.50	30.26	30.09	30.85	30.76
Undifferentiated Services (ISIC Rev.4 Code: 97, 98)	10.80	10.62	10.37	10.64	10.49	10.36	10.12	10.01	9.63	9.74	10.28
All Services (Average)	6.94	6.66	6.50	6.83	6.93	6.80	6.68	6.79	6.95	6.92	7.02
Business Sector Services (Average)	6.96	6.77	6.61	6.97	7.05	6.92	6.81	6.94	7.14	7.11	7.22
Business Sector Services (Average excluding real estate)	6.84	6.68	6.52	6.89	7.00	6.90	6.78	6.88	7.07	7.02	7.11
All Industries (Average)	5.83	5.80	5.77	5.99	5.93	5.66	5.54	5.50	5.35	5.25	5.07

Note: DGM (Domestic Global Market) Ratio is computed by dividing gross productive output of a country by the OECD productive output in the corresponding service sector. In this computation the OECD output is taken as a proxy for global market output. Therefore, it is likely that these numbers are inflated (i.e., likely to be lower when the total global market output is factored in). Gross productive output data is secured from OECD databases.

Table 4.53 Domestic global market ratio: Hungary

Industry	2000	2001	2002	2003	2004	2005	2006	2007	2008	2009	2010
Construction (ISIC Rev.4 Code: 41)	0.20	0.24	0.31	0.33	0.36	0.38	0.35	0.36	0.40	0.37	0.35
Automotive Retail & Repair (ISIC Rev.4 Code: 45)	0.53	0.65	0.76	0.81	0.86	0.86	0.86	0.95	1.06	0.91	0.86
Wholesale Trade (ISIC Rev.4 Code: 46)	0.53	0.65	0.78	0.81	0.81	0.89	0.90	0.93	1.08	0.87	0.82
Retail Trade (ISIC Rev.4 Code: 47)	0.51	0.62	0.75	0.80	0.88	0.92	0.91	0.99	1.10	0.96	0.91
Land Transport (ISIC Rev.4 Code: 49)	0.37	0.41	0.48	0.51	0.58	0.60	0.60	0.70	0.71	0.65	0.61
Water Transport (ISIC Rev.4 Code: 50)	0.02	0.03	0.03	0.03	0.03	0.03	0.03	0.03	0.04	0.04	0.04
Air Transport (ISIC Rev.4 Code: 51)	0.13	0.14	0.17	0.18	0.23	0.24	0.24	0.29	0.31	0.31	0.31
Support Activities for Transport (ISIC Rev.4 Code: 52)	0.19	0.21	0.23	0.26	0.28	0.31	0.32	0.39	0.57	0.55	0.54
Postal & Courier Activities (ISIC Rev.4 Code: 53)	0.18	0.27	0.33	0.29	0.35	0.38	0.37	0.41	0.46	0.40	0.38
Hotel & Restaurants (ISIC Rev.4 Code: 55, 56)	0.16	0.17	0.22	0.24	0.27	0.27	0.25	0.29	0.32	0.28	0.25
Publishing Activities (ISIC Rev.4 Code: 58)	0.12	0.13	0.17	0.18	0.21	0.21	0.20	0.21	0.21	0.17	0.18
Audiovisual & Broadcasting (ISIC Rev.4 Code: 59, 60)	0.38	0.50	0.67	0.78	0.82	0.84	0.79	0.89	0.88	1.20	1.19
Telecommunications (ISIC Rev.4 Code: 61)	0.11	0.19	0.22	0.27	0.30	0.32	0.31	0.34	0.35	0.29	0.27
IT & Information Services (ISIC Rev.4 Code: 62, 63)	0.13	0.15	0.19	0.23	0.26	0.26	0.27	0.31	0.34	0.33	0.31

Financial Service Activities (ISIC Rev.4 Code: 64)	0.14	0.15	0.18	0.22	0.25	0.28	0.28	0.32	0.32	0.28	0.27
Insurance & Pensions (ISIC Rev.4 Code: 65)	0.08	0.08	0.10	0.11	0.14	0.14	0.13	0.14	0.17	0.16	0.14
Auxiliary Financial Services (ISIC Rev.4 Code: 66)	0.04	0.05	0.07	0.09	0.09	0.11	0.08	0.10	0.13	0.10	0.11
Legal, Accounting & Consultancy (ISIC Rev.4 Code: 69,70)	0.30	0.36	0.44	0.48	0.52	0.50	0.51	0.55	0.61	0.52	0.48
Architectural & Technical Services (ISIC Rev.4 Code: 71)	0.41	0.51	0.62	0.70	0.79	0.74	0.69	0.66	0.70	0.65	0.70
Research & Development (ISIC Rev.4 Code: 72)	0.33	0.40	0.43	0.50	0.54	0.60	0.58	0.58	0.56	0.57	0.53
Advertising & Market Research (ISIC Rev.4 Code: 73)	0.23	0.30	0.39	0.47	0.45	0.58	0.51	0.56	0.65	0.55	0.55
Other Professionals Services (ISIC Rev.4 Code: 74, 75)	0.45	0.62	0.75	0.88	0.98	1.04	1.04	1.07	1.01	1.00	0.99
Rental & Leasing Activities (ISIC Rev.4 Code: 77)	0.07	0.08	0.11	0.10	0.12	0.13	0.13	0.14	0.15	0.18	0.16
Employment Activities (ISIC Rev.4 Code: 78)	0.12	0.15	0.19	0.24	0.33	0.40	0.42	0.44	0.45	0.48	0.46
Travel Service (ISIC Rev.4 Code: 79)	0.22	0.27	0.33	0.45	0.52	0.52	0.54	0.55	0.55	0.22	0.21
Security & Building Services (ISIC Rev.4 Code: 80, 81, 82)	0.52	0.67	0.79	0.91	0.95	0.90	0.83	0.88	1.02	0.91	0.84
Public Administration & Defense (ISIC Rev.4 Code: 84)	0.17	0.18	0.22	0.25	0.27	0.27	0.28	0.31	0.33	0.28	0.26
Education (ISIC Rev.4 Code: 85)	0.16	0.18	0.22	0.28	0.30	0.31	0.29	0.30	0.33	0.28	0.26

(continued)

Table 4.53 (continued)

Industry	2000	2001	2002	2003	2004	2005	2006	2007	2008	2009	2010
Human Health Activities (ISIC Rev.4 Code: 86)	0.40	0.44	0.53	0.60	0.60	0.61	0.55	0.57	0.59	0.53	0.49
Residential Care & Social Work (ISIC Rev.4 Code: 87, 88)	0.20	0.24	0.29	0.34	0.38	0.40	0.37	0.40	0.42	0.35	0.31
Creative Entertainment Activities (ISIC Rev.4 Code: 90, 91, 92)	0.63	0.73	0.96	0.91	0.86	0.96	0.88	0.96	0.98	0.84	0.90
Sports & Recreation Activities (ISIC Rev.4 Code: 93)	0.59	0.60	0.70	0.74	0.76	0.73	0.67	0.69	0.81	0.69	0.60
Membership Organizations (ISIC Rev.4 Code: 94)	0.48	0.56	0.65	0.66	0.77	0.85	0.79	0.88	0.90	0.80	0.74
Repair Services (ISIC Rev.4 Code: 95)	0.52	0.59	0.75	0.83	0.73	0.90	0.83	0.81	0.94	1.18	1.07
Personal Service Activities (ISIC Rev.4 Code: 96)	0.46	0.56	0.63	0.65	0.75	0.75	0.73	0.80	0.92	0.75	0.73
Undifferentiated Services (ISIC Rev.4 Code: 97, 98)	0.02	0.02	0.03	0.02	0.02	0.02	0.02	0.02	0.02	0.02	0.03
All Services (Average)	0.15	0.17	0.21	0.24	0.27	0.28	0.27	0.30	0.33	0.29	0.27
Business Sector Services (Average)	0.15	0.18	0.21	0.24	0.27	0.28	0.27	0.31	0.34	0.30	0.29
Business Sector Services (Average excluding real estate)	0.16	0.18	0.22	0.25	0.28	0.29	0.28	0.31	0.36	0.31	0.29
All Industries (Average)	0.17	0.19	0.23	0.26	0.27	0.28	0.27	0.28	0.29	0.25	0.23

Note: DGM (Domestic Global Market) Ratio is computed by dividing gross productive output of a country by the OECD productive output in the corresponding service sector. In this computation the OECD output is taken as a proxy for global market output. Therefore, it is likely that these numbers are inflated (i.e., likely to be lower when the total global market output is factored in). Gross productive output data is secured from OECD databases.

Table 4.54 Domestic global market ratio: Italy

Industry	2000	2001	2002	2003	2004	2005	2006	2007	2008	2009	2010
Construction (ISIC Rev.4 Code: 41)	5.74	5.73	6.05	6.43	6.70	6.65	6.52	6.73	6.91	6.92	7.13
Automotive Retail & Repair (ISIC Rev.4 Code: 45)	15.18	14.96	14.55	14.29	14.59	14.64	14.77	14.72	14.68	14.99	15.20
Wholesale Trade (ISIC Rev.4 Code: 46)	14.81	14.66	14.57	14.41	14.23	13.82	13.73	13.32	12.88	13.09	12.93
Retail Trade (ISIC Rev.4 Code: 47)	13.83	14.13	13.95	13.94	13.90	13.74	13.53	13.45	13.22	13.07	13.56
Land Transport (ISIC Rev.4 Code: 49)	9.32	9.18	9.38	10.23	10.53	10.52	10.35	10.63	10.69	11.28	11.23
Water Transport (ISIC Rev.4 Code: 50)	5.84	5.70	5.90	5.77	5.33	5.27	5.11	4.90	4.64	4.63	4.75
Air Transport (ISIC Rev.4 Code: 51)	3.25	3.59	3.17	3.30	3.21	3.13	3.07	2.94	2.72	2.29	2.35
Support Activities for Transport (ISIC Rev.4 Code: 52)	8.17	8.20	8.52	8.78	8.82	8.65	8.35	8.20	7.85	7.83	7.86
Postal & Courier Activities (ISIC Rev.4 Code: 53)	3.74	4.00	3.70	3.82	3.74	3.83	3.74	4.35	4.05	3.86	4.16
Hotel & Restaurants (ISIC Rev.4 Code: 55, 56)	6.51	6.39	6.31	6.75	6.98	6.85	6.84	7.02	7.21	7.19	7.19
Publishing Activities (ISIC Rev.4 Code: 58)	2.18	2.18	2.15	2.33	2.44	2.43	2.40	2.37	2.32	2.09	2.16
Audiovisual & Broadcasting (ISIC Rev.4 Code: 59, 60)	12.79	13.18	12.83	12.70	12.62	12.15	12.04	12.46	12.47	12.43	12.53
Telecommunications (ISIC Rev.4 Code: 61)	1.84	3.33	3.53	3.83	4.12	4.18	4.14	4.13	4.06	4.00	3.95

(continued)

Table 4.54 (continued)

Industry	2000	2001	2002	2003	2004	2005	2006	2007	2008	2009	2010
IT & Information Services (ISIC Rev.4 Code: 62, 63)	4.78	4.91	5.67	5.95	5.99	5.73	5.56	5.46	5.29	5.30	5.21
Financial Service Activities (ISIC Rev.4 Code: 64)	4.54	4.14	3.84	4.12	4.37	4.31	4.50	4.75	4.88	4.54	4.60
Insurance & Pensions (ISIC Rev.4 Code: 65)	1.35	1.74	1.71	1.85	1.71	1.96	1.88	2.20	2.08	1.64	1.78
Auxiliary Financial Services (ISIC Rev.4 Code: 66)	1.44	1.62	1.90	2.08	2.18	2.04	1.83	1.86	2.21	2.24	1.98
Legal, Accounting & Consultancy (ISIC Rev.4 Code: 69,70)	12.57	12.04	11.80	12.05	12.03	11.71	10.87	10.46	10.30	11.07	11.27
Architectural & Technical Services (ISIC Rev.4 Code: 71)	9.59	9.35	9.66	9.90	10.23	9.62	9.11	8.50	7.73	8.08	8.61
Research & Development (ISIC Rev.4 Code: 72)	9.71	10.06	10.07	10.53	10.28	10.44	10.37	10.10	9.67	9.33	9.19
Advertising & Market Research (ISIC Rev.4 Code: 73)	11.35	12.30	12.38	12.47	12.54	12.23	12.07	11.90	11.71	12.43	12.38
Other Professionals Services (ISIC Rev.4 Code: 74, 75)	23.54	23.19	22.49	23.26	23.56	23.09	22.44	21.64	21.78	22.21	22.57
Rental & Leasing Activities (ISIC Rev.4 Code: 77)	1.41	1.56	1.76	1.87	1.88	1.96	1.92	2.03	2.01	2.20	2.18
Employment Activities (ISIC Rev.4 Code: 78)	2.87	4.13	4.83	5.44	5.94	5.92	5.95	5.88	5.59	4.02	4.12
Travel Service (ISIC Rev.4 Code: 79)	14.54	14.47	14.76	14.70	13.89	14.34	14.76	14.61	13.32	12.19	12.43
Security & Building Services (ISIC Rev.4 Code: 80, 81, 82)	13.78	13.97	13.84	13.95	13.95	13.61	13.32	12.84	12.37	12.73	13.02
Public Administration & Defense (ISIC Rev.4 Code: 84)	3.06	2.94	2.91	3.23	3.39	3.38	3.31	3.32	3.35	3.28	3.23

Education (ISIC Rev.4 Code: 85)	3.72	3.55	3.61	3.86	3.86	3.86	3.79	3.85	3.69	3.64	3.60
Human Health Activities (ISIC Rev.4 Code: 86)	11.36	11.41	11.15	10.99	11.38	11.58	11.55	11.14	11.52	11.40	11.27
Residential Care & Social Work (ISIC Rev.4 Code: 87, 88)	4.82	4.81	4.77	4.86	5.07	5.13	5.22	5.08	5.16	5.13	5.19
Creative Entertainment Activities (ISIC Rev.4 Code: 90, 91, 92)	10.66	10.95	11.20	11.28	11.24	10.90	10.78	10.84	10.86	11.13	11.63
Sports & Recreation Activities (ISIC Rev.4 Code: 93)	10.18	10.12	9.73	9.53	9.61	9.02	8.95	9.05	9.00	9.18	9.40
Membership Organizations (ISIC Rev.4 Code: 94)	4.56	4.47	4.47	4.37	4.52	4.64	4.69	4.62	4.51	4.55	4.49
Repair Services (ISIC Rev.4 Code: 95)	19.85	19.33	18.71	17.99	18.02	17.17	16.87	16.94	16.29	15.65	16.17
Personal Service Activities (ISIC Rev.4 Code: 96)	11.33	11.41	11.19	11.24	11.07	11.07	11.01	10.63	10.83	11.00	11.18
Undifferentiated Services (ISIC Rev.4 Code: 97, 98)	16.91	17.31	18.13	18.32	18.81	19.47	19.62	20.27	21.09	21.96	23.01
All Services (Average)	4.15	4.10	4.13	4.43	4.58	4.53	4.48	4.55	4.62	4.54	4.53
Business Sector Services (Average)	4.43	4.42	4.48	4.80	4.94	4.85	4.79	4.88	4.96	4.91	4.89
Business Sector Services (Average excluding real estate)	4.62	4.63	4.67	4.98	5.10	5.02	4.93	5.01	5.07	5.01	4.94
All Industries (Average)	3.73	3.78	3.91	4.10	4.11	3.94	3.85	3.80	3.66	3.51	3.29

Note: DGM (Domestic Global Market) Ratio is computed by dividing gross productive output of a country by the OECD productive output in the corresponding service sector. In this computation the OECD output is taken as a proxy for global market output. Therefore, it is likely that these numbers are inflated (i.e., likely to be lower when the total global market output is factored in). Gross productive output data is secured from OECD databases.

Table 4.55 Domestic global market ratio: Korea

Industry	2000	2001	2002	2003	2004	2005	2006	2007	2008	2009	2010
Construction (ISIC Rev.4 Code: 41)	3.11	2.95	3.29	3.64	3.71	3.84	3.92	4.10	3.67	3.73	4.24
Automotive Retail & Repair (ISIC Rev.4 Code: 45)											
Wholesale Trade (ISIC Rev.4 Code: 46)											
Retail Trade (ISIC Rev.4 Code: 47)											
Land Transport (ISIC Rev.4 Code: 49)											
Water Transport (ISIC Rev.4 Code: 50)											
Air Transport (ISIC Rev.4 Code: 51)											
Support Activities for Transport (ISIC Rev.4 Code: 52)											
Postal & Courier Activities (ISIC Rev.4 Code: 53)											
Hotel & Restaurants (ISIC Rev.4 Code: 55, 56)	2.82	2.71	2.99	3.01	2.94	3.17	3.39	3.43	3.13	3.01	3.29
Publishing Activities (ISIC Rev.4 Code: 58)	2.59	2.42	3.01	3.10	3.13	3.60	4.00	3.93	3.22	2.97	3.31
Audiovisual & Broadcasting (ISIC Rev.4 Code: 59, 60)	4.16	4.09	5.28	5.17	5.14	5.89	6.49	6.47	5.21	4.81	5.68
Telecommunications (ISIC Rev.4 Code: 61)	1.42	2.27	2.49	2.56	2.65	2.96	3.16	3.25	2.78	2.50	2.72
IT & Information Services (ISIC Rev.4 Code: 62, 63)	0.43	0.45	0.53	0.57	0.68	0.80	0.89	0.88	0.75	0.83	0.95

Financial Service Activities (ISIC Rev.4 Code: 64)											
Insurance & Pensions (ISIC Rev.4 Code: 65)											
Auxiliary Financial Services (ISIC Rev.4 Code: 66)											
Legal, Accounting & Consultancy (ISIC Rev.4 Code: 69,70)											
Architectural & Technical Services (ISIC Rev.4 Code: 71)											
Research & Development (ISIC Rev.4 Code: 72)	3.06	2.99	3.15	3.16	3.21	3.77	4.11	4.15	3.56	3.59	4.07
Advertising & Market Research (ISIC Rev.4 Code: 73)	4.07	3.57	4.40	4.09	3.75	4.10	4.48	4.35	3.45	3.28	4.01
Other Professionals Services (ISIC Rev.4 Code: 74, 75)											
Rental & Leasing Activities (ISIC Rev.4 Code: 77)	0.29	0.19	0.23	0.23	0.15	0.19	0.18	0.24	0.18	0.18	0.24
Employment Activities (ISIC Rev.4 Code: 78)											
Travel Service (ISIC Rev.4 Code: 79)											
Security & Building Services (ISIC Rev.4 Code: 80, 81, 82)											
Public Administration & Defense (ISIC Rev.4 Code: 84)	1.40	1.30	1.38	1.44	1.51	1.72	1.89	1.95	1.66	1.51	1.62
Education (ISIC Rev.4 Code: 85)	1.98	1.91	2.07	2.24	2.39	2.69	3.00	3.10	2.74	2.53	2.77

(continued)

Table 4.55 (continued)

Industry	2000	2001	2002	2003	2004	2005	2006	2007	2008	2009	2010
Human Health Activities (ISIC Rev.4 Code: 86)	3.59	4.06	4.28	4.12	4.13	4.80	5.37	5.54	4.72	4.58	5.18
Residential Care & Social Work (ISIC Rev.4 Code: 87, 88)	1.39	1.38	1.48	1.48	1.54	1.75	1.89	2.05	1.72	1.61	1.77
Creative Entertainment Activities (ISIC Rev.4 Code: 90, 91, 92)	2.09	2.03	2.17	2.11	2.12	2.46	2.78	2.82	2.49	2.34	2.71
Sports & Recreation Activities (ISIC Rev.4 Code: 93)	8.83	9.38	11.15	10.53	10.30	11.30	11.75	12.00	10.23	9.67	10.88
Membership Organizations (ISIC Rev.4 Code: 94)	4.28	4.21	4.68	4.47	4.33	5.06	5.51	5.59	4.79	4.26	4.84
Repair Services (ISIC Rev.4 Code: 95)											
Personal Service Activities (ISIC Rev.4 Code: 96)	4.60	4.52	5.07	4.79	4.49	4.93	5.30	5.30	4.75	4.49	4.96
Undifferentiated Services (ISIC Rev.4 Code: 97, 98)											
All Services (Average)	1.60	1.53	1.69	1.72	1.75	1.93	2.08	2.16	1.91	1.77	2.01
Business Sector Services (Average)	1.67	1.59	1.76	1.78	1.79	1.96	2.09	2.16	1.92	1.80	2.05
Business Sector Services (Average excluding real estate)	1.68	1.61	1.81	1.83	1.84	2.02	2.16	2.24	2.00	1.89	2.15
All Industries (Average)	1.90	1.82	2.03	2.07	2.14	2.31	2.40	2.40	2.12	2.04	2.22

Note: DGM (Domestic Global Market) Ratio is computed by dividing gross productive output of a country by the OECD productive output in the corresponding service sector. In this computation the OECD output is taken as a proxy for global market output. Therefore, it is likely that these numbers are inflated (i.e., likely to be lower when the total global market output is factored in). Gross productive output data is secured from OECD databases.

Table 4.56 Domestic global market ratio: Netherlands

Industry	2000	2001	2002	2003	2004	2005	2006	2007	2008	2009	2010
Construction (ISIC Rev.4 Code: 41)	1.68	1.71	1.65	1.68	1.64	1.60	1.64	1.80	2.20	2.00	
Automotive Retail & Repair (ISIC Rev.4 Code: 45)	4.25	4.06	4.16	4.21	4.35	4.23	4.18	4.32	4.92	4.16	
Wholesale Trade (ISIC Rev.4 Code: 46)	4.32	4.28	4.43	4.46	4.78	4.78	4.78	5.06	5.88	4.60	
Retail Trade (ISIC Rev.4 Code: 47)	3.04	2.99	3.09	3.28	3.35	3.30	3.32	3.39	3.53	3.16	
Land Transport (ISIC Rev.4 Code: 49)	1.35	1.32	1.32	1.38	1.40	1.36	1.37	1.41	1.74	1.38	
Water Transport (ISIC Rev.4 Code: 50)	2.72	2.94	2.73	2.84	2.87	3.01	2.35	2.38	3.17	2.13	
Air Transport (ISIC Rev.4 Code: 51)	2.21	2.15	2.07	2.01	2.08	2.15	2.17	2.07	2.73	1.83	
Support Activities for Transport (ISIC Rev.4 Code: 52)	1.15	1.13	1.14	1.23	1.32	1.34	1.32	1.36	1.68	1.34	
Postal & Courier Activities (ISIC Rev.4 Code: 53)	1.51	1.49	1.49	1.63	1.83	1.79	1.77	1.81	1.96	1.71	
Hotel & Restaurants (ISIC Rev.4 Code: 55, 56)	0.93	0.90	0.88	0.88	0.89	0.86	0.85	0.88	0.93	0.78	
Publishing Activities (ISIC Rev.4 Code: 58)	0.94	0.91	0.91	0.95	0.98	0.96	0.92	0.95	1.02	0.88	
Audiovisual & Broadcasting (ISIC Rev.4 Code: 59, 60)	2.62	2.71	2.90	3.17	3.33	3.19	3.19	3.34	3.44	2.92	
Telecommunications (ISIC Rev.4 Code: 61)	1.00	1.08	1.16	1.30	1.29	1.27	1.21	1.23	1.22	1.02	

(continued)

Table 4.56 (continued)

Industry	2000	2001	2002	2003	2004	2005	2006	2007	2008	2009	2010
IT & Information Services (ISIC Rev.4 Code: 62, 63)	1.44	1.55	1.48	1.52	1.57	1.60	1.59	1.67	1.86	1.49	
Financial Service Activities (ISIC Rev.4 Code: 64)	1.46	1.36	1.40	1.58	1.63	1.65	1.44	1.33	1.53	1.78	
Insurance & Pensions (ISIC Rev.4 Code: 65)	1.17	1.14	1.04	1.24	1.27	1.37	1.38	1.38	1.33	1.21	
Auxiliary Financial Services (ISIC Rev.4 Code: 66)	0.60	0.63	0.58	0.58	0.56	0.46	0.47	0.57	0.55	0.43	
Legal, Accounting & Consultancy (ISIC Rev.4 Code: 69,70)	4.05	4.12	3.90	4.16	4.31	4.25	4.15	4.26	4.98	4.31	
Architectural & Technical Services (ISIC Rev.4 Code: 71)	3.88	4.01	3.99	4.20	4.16	4.10	4.02	4.07	4.79	4.42	
Research & Development (ISIC Rev.4 Code: 72)	2.40	2.37	2.28	2.56	2.90	2.87	2.83	2.68	2.72	2.24	
Advertising & Market Research (ISIC Rev.4 Code: 73)	4.18	4.09	4.16	4.23	4.44	4.36	4.22	4.19	4.88	3.92	
Other Professionals Services (ISIC Rev.4 Code: 74, 75)	3.57	3.75	3.80	3.81	4.16	4.11	4.08	4.34	5.30	4.67	
Rental & Leasing Activities (ISIC Rev.4 Code: 77)	1.11	1.20	1.05	1.07	1.08	1.08	1.05	1.08	1.30	1.15	
Employment Activities (ISIC Rev.4 Code: 78)	9.74	10.18	9.57	9.30	9.13	8.77	9.08	10.45	14.40	10.72	
Travel Service (ISIC Rev.4 Code: 79)	3.95	4.50	4.76	4.97	4.97	4.97	4.86	5.15	5.94	5.03	
Security & Building Services (ISIC Rev.4 Code: 80, 81, 82)	2.46	2.55	2.58	2.72	2.78	2.74	2.70	2.77	3.12	2.77	
Public Administration & Defense (ISIC Rev.4 Code: 84)	1.08	1.07	1.09	1.21	1.22	1.19	1.17	1.15	1.22	1.13	

Education (ISIC Rev.4 Code: 85)	0.97	0.97	1.00	1.10	1.14	1.13	1.08	1.10	1.18	1.09
Human Health Activities (ISIC Rev.4 Code: 86)	2.39	2.43	2.64	2.97	3.17	3.11	3.06	3.11	3.27	3.03
Residential Care & Social Work (ISIC Rev.4 Code: 87, 88)	5.16	5.04	5.40	6.08	6.20	6.14	5.99	5.91	6.17	5.65
Creative Entertainment Activities (ISIC Rev.4 Code: 90, 91, 92)	3.43	3.49	3.62	3.91	4.13	4.03	4.02	4.11	4.31	3.84
Sports & Recreation Activities (ISIC Rev.4 Code: 93)	2.38	2.27	2.30	2.56	2.62	2.57	2.46	2.46	2.59	2.30
Membership Organizations (ISIC Rev.4 Code: 94)	2.69	2.63	2.66	2.90	3.13	3.15	3.11	3.19	3.30	3.04
Repair Services (ISIC Rev.4 Code: 95)	1.77	1.86	1.88	1.98	2.02	1.94	1.99	2.16	2.51	2.13
Personal Service Activities (ISIC Rev.4 Code: 96)	1.66	1.69	1.75	1.90	1.97	1.96	1.88	1.90	1.99	1.79
Undifferentiated Services (ISIC Rev.4 Code: 97, 98)	2.12	2.15	2.21	2.40	2.56	2.51	2.40	2.35	2.53	2.43
All Services (Average)	1.18	1.17	1.17	1.24	1.27	1.26	1.23	1.26	1.39	1.21
Business Sector Services (Average)	1.23	1.22	1.21	1.26	1.29	1.28	1.25	1.30	1.45	1.23
Business Sector Services (Average excluding real estate)	1.31	1.31	1.29	1.35	1.38	1.36	1.33	1.38	1.56	1.30
All Industries (Average)	1.27	1.26	1.24	1.30	1.33	1.32	1.31	1.35	1.58	1.28

Note: DGM (Domestic Global Market) Ratio is computed by dividing gross productive output of a country by the OECD productive output in the corresponding service sector. In this computation the OECD output is taken as a proxy for global market output. Therefore, it is likely that these numbers are inflated (i.e., likely to be lower when the total global market output is factored in). Gross productive output data is secured from OECD databases.

Table 4.57 Domestic global market ratio: Norway

Industry	2000	2001	2002	2003	2004	2005	2006	2007	2008	2009	2010
Construction (ISIC Rev.4 Code: 41)	0.63	0.67	0.78	0.85	0.91	0.99	1.08	1.34	1.48	1.35	1.48
Automotive Retail & Repair (ISIC Rev.4 Code: 45)	1.92	1.77	1.95	1.92	1.85	1.98	1.88	2.42	2.74	2.48	
Wholesale Trade (ISIC Rev.4 Code: 46)	1.39	1.48	1.56	1.60	1.64	1.68	1.79	1.93	2.18	1.90	
Retail Trade (ISIC Rev.4 Code: 47)	1.36	1.38	1.66	1.67	1.68	1.86	1.95	2.05	2.03	1.87	
Land Transport (ISIC Rev.4 Code: 49)	0.84	0.84	0.93	0.98	0.97	1.00	1.01	1.07	1.15	1.15	1.12
Water Transport (ISIC Rev.4 Code: 50)	8.19	9.20	9.46	9.46	9.09	9.20	8.81	8.51	8.17	8.09	7.94
Air Transport (ISIC Rev.4 Code: 51)	0.86	0.92	1.06	1.01	1.02	1.09	1.12	1.21	1.31	1.34	
Support Activities for Transport (ISIC Rev.4 Code: 52)	0.94	1.03	1.13	1.16	1.17	1.21	1.28	1.46	1.53	1.49	
Postal & Courier Activities (ISIC Rev.4 Code: 53)	0.75	0.77	0.83	0.90	0.94	1.03	0.93	1.01	1.07	1.06	1.10
Hotel & Restaurants (ISIC Rev.4 Code: 55, 56)	0.39	0.39	0.43	0.45	0.44	0.45	0.46	0.51	0.55	0.52	0.53
Publishing Activities (ISIC Rev.4 Code: 58)	0.64	0.68	0.74	0.81	0.87	0.91	0.93	1.00	1.02	0.95	
Audiovisual & Broadcasting (ISIC Rev.4 Code: 59, 60)	0.95	0.97	1.12	1.16	1.18	1.24	1.29	1.31	1.52	1.36	
Telecommunications (ISIC Rev.4 Code: 61)	0.32	0.48	0.58	0.66	0.68	0.65	0.68	0.73	0.77	0.72	0.74
IT & Information Services (ISIC Rev.4 Code: 62, 63)	0.55	0.60	0.64	0.64	0.69	0.75	0.77	0.84	0.91	0.85	0.84

Activity											
Financial Service Activities (ISIC Rev.4 Code: 64)	0.44	0.44	0.46	0.55	0.59	0.59	0.56	0.63	0.66	0.69	
Insurance & Pensions (ISIC Rev.4 Code: 65)	0.13	0.12	0.17	0.24	0.29	0.32	0.30	0.38	0.41	0.36	
Auxiliary Financial Services (ISIC Rev.4 Code: 66)	0.10	0.10	0.12	0.13	0.15	0.19	0.29	0.40	0.35	0.30	
Legal, Accounting & Consultancy (ISIC Rev.4 Code: 69,70)	0.62	0.63	0.71	0.68	0.68	0.72	0.91	1.00	0.90	0.91	
Architectural & Technical Services (ISIC Rev.4 Code: 71)	2.21	2.49	2.57	2.47	2.57	2.78	3.27	3.78	4.33	4.34	
Research & Development (ISIC Rev.4 Code: 72)	0.87	0.85	0.93	0.94	1.14	1.06	1.04	1.07	1.18	1.10	1.16
Advertising & Market Research (ISIC Rev.4 Code: 73)	1.21	1.24	1.36	1.38	1.43	1.53	0.98	1.03	1.67	1.52	
Other Professionals Services (ISIC Rev.4 Code: 74, 75)	1.14	1.16	1.17	1.24	1.26	1.42	1.53	1.72	1.96	1.45	
Rental & Leasing Activities (ISIC Rev.4 Code: 77)	0.31	0.33	0.41	0.38	0.34	0.37	0.39	0.48	0.54	0.58	
Employment Activities (ISIC Rev.4 Code: 78)	0.86	1.04	1.29	1.29	1.29	1.54	1.88	2.09	2.29	2.37	
Travel Service (ISIC Rev.4 Code: 79)	1.52	1.59	1.81	1.75	1.73	1.67	1.75	1.93	2.10	2.09	
Security & Building Services (ISIC Rev.4 Code: 80, 81, 82)	1.27	1.27	1.30	1.34	1.31	1.43	1.47	1.59	1.64	1.60	
Public Administration & Defense (ISIC Rev.4 Code: 84)	0.51	0.52	0.55	0.56	0.56	0.55	0.56	0.65	0.69	0.67	0.69
Education (ISIC Rev.4 Code: 85)	0.53	0.56	0.64	0.70	0.72	0.74	0.74	0.78	0.83	0.81	0.85

(continued)

Table 4.57 (continued)

Industry	2000	2001	2002	2003	2004	2005	2006	2007	2008	2009	2010
Human Health Activities (ISIC Rev.4 Code: 86)	1.32	1.42	1.59	1.63	1.59	1.64	1.64	1.78	1.86	1.77	1.81
Residential Care & Social Work (ISIC Rev.4 Code: 87, 88)	3.21	3.51	3.96	4.02	3.98	4.14	4.19	4.45	4.68	4.56	4.66
Creative Entertainment Activities (ISIC Rev.4 Code: 90, 91, 92)	2.05	2.01	2.37	2.49	2.61	2.66	2.59	2.23	2.20	2.17	
Sports & Recreation Activities (ISIC Rev.4 Code: 93)	1.25	1.27	1.37	1.41	1.36	1.42	1.40	1.63	1.73	1.72	
Membership Organizations (ISIC Rev.4 Code: 94)	1.37	1.46	1.66	1.67	1.70	1.74	1.73	1.89	2.06	2.04	
Repair Services (ISIC Rev.4 Code: 95)	0.70	0.77	0.82	0.76	0.74	0.75	0.72	0.85	0.84	0.85	
Personal Service Activities (ISIC Rev.4 Code: 96)	0.66	0.71	0.83	0.85	0.82	0.85	0.92	1.00	1.06	1.01	
Undifferentiated Services (ISIC Rev.4 Code: 97, 98)	0.28	0.30	0.33	0.17	0.13	0.12	0.09	0.06	0.07	0.07	0.08
All Services (Average)	0.53	0.54	0.60	0.64	0.66	0.69	0.70	0.78	0.85	0.80	0.82
Business Sector Services (Average)	0.52	0.54	0.59	0.63	0.65	0.68	0.70	0.78	0.86	0.80	0.81
Business Sector Services (Average excluding real estate)	0.53	0.55	0.61	0.65	0.67	0.70	0.72	0.81	0.89	0.82	0.83
All Industries (Average)	0.45	0.48	0.53	0.56	0.57	0.60	0.62	0.65	0.69	0.63	0.61

Note: DGM (Domestic Global Market) Ratio is computed by dividing gross productive output of a country by the OECD productive output in the corresponding service sector. In this computation the OECD output is taken as a proxy for global market output. Therefore, it is likely that these numbers are inflated (i.e., likely to be lower when the total global market output is factored in). Gross productive output data is secured from OECD databases.

Table 4.58 Domestic global market ratio: Slovenia

Industry	2000	2001	2002	2003	2004	2005	2006	2007	2008	2009	2010
Construction (ISIC Rev.4 Code: 41)								0.25	0.29	0.26	0.23
Automotive Retail & Repair (ISIC Rev.4 Code: 45)								0.31	0.34	0.33	0.33
Wholesale Trade (ISIC Rev.4 Code: 46)								0.31	0.33	0.33	0.30
Retail Trade (ISIC Rev.4 Code: 47)								0.35	0.36	0.37	0.38
Land Transport (ISIC Rev.4 Code: 49)								0.22	0.24	0.24	0.24
Water Transport (ISIC Rev.4 Code: 50)								0.10	0.12	0.10	0.10
Air Transport (ISIC Rev.4 Code: 51)								0.07	0.08	0.08	0.07
Support Activities for Transport (ISIC Rev.4 Code: 52)								0.17	0.18	0.17	0.20
Postal & Courier Activities (ISIC Rev.4 Code: 53)								0.12	0.14	0.14	0.15
Hotel & Restaurants (ISIC Rev.4 Code: 55, 56)								0.09	0.10	0.10	0.09
Publishing Activities (ISIC Rev.4 Code: 58)								0.06	0.07	0.06	0.06
Audiovisual & Broadcasting (ISIC Rev.4 Code: 59, 60)								0.19	0.21	0.20	0.20
Telecommunications (ISIC Rev.4 Code: 61)								0.10	0.12	0.11	0.11

(continued)

Table 4.58 (continued)

Industry	2000	2001	2002	2003	2004	2005	2006	2007	2008	2009	2010
IT & Information Services (ISIC Rev.4 Code: 62, 63)								0.09	0.10	0.10	0.10
Financial Service Activities (ISIC Rev.4 Code: 64)								0.09	0.10	0.09	0.10
Insurance & Pensions (ISIC Rev.4 Code: 65)								0.06	0.06	0.07	0.08
Auxiliary Financial Services (ISIC Rev.4 Code: 66)								0.02	0.02	0.02	0.01
Legal, Accounting & Consultancy (ISIC Rev.4 Code: 69,70)								0.15	0.16	0.18	0.19
Architectural & Technical Services (ISIC Rev.4 Code: 71)								0.53	0.57	0.52	0.56
Research & Development (ISIC Rev.4 Code: 72)								0.23	0.25	0.25	0.27
Advertising & Market Research (ISIC Rev.4 Code: 73)								0.25	0.28	0.29	0.30
Other Professionals Services (ISIC Rev.4 Code: 74, 75)								0.47	0.51	0.52	0.51
Rental & Leasing Activities (ISIC Rev.4 Code: 77)								0.01	0.01	0.01	0.01
Employment Activities (ISIC Rev.4 Code: 78)								0.33	0.32	0.34	0.34
Travel Service (ISIC Rev.4 Code: 79)								0.41	0.44	0.42	0.41
Security & Building Services (ISIC Rev.4 Code: 80, 81, 82)								0.12	0.13	0.13	0.14
Public Administration & Defense (ISIC Rev.4 Code: 84)								0.07	0.08	0.07	0.08

Education (ISIC Rev.4 Code: 85)	0.10	0.11	0.11	0.11
Human Health Activities (ISIC Rev.4 Code: 86)	0.20	0.21	0.22	0.21
Residential Care & Social Work (ISIC Rev.4 Code: 87, 88)	0.10	0.11	0.12	0.12
Creative Entertainment Activities (ISIC Rev.4 Code: 90, 91, 92)	0.43	0.45	0.45	0.44
Sports & Recreation Activities (ISIC Rev.4 Code: 93)	0.19	0.20	0.21	0.20
Membership Organizations (ISIC Rev.4 Code: 94)	0.25	0.26	0.25	0.24
Repair Services (ISIC Rev.4 Code: 95)	0.21	0.21	0.20	0.20
Personal Service Activities (ISIC Rev.4 Code: 96)	0.16	0.17	0.17	0.18
Undifferentiated Services (ISIC Rev.4 Code: 97, 98)	0.03	0.03	0.03	0.04
All Services (Average)	0.09	0.10	0.10	0.10
Business Sector Services (Average)	0.10	0.11	0.10	0.10
Business Sector Services (Average excluding real estate)	0.11	0.12	0.11	0.11
All Industries (Average)	0.09	0.09	0.08	0.08

Note: DGM (Domestic Global Market) Ratio is computed by dividing gross productive output of a country by the OECD productive output in the corresponding service sector. In this computation the OECD output is taken as a proxy for global market output. Therefore, it is likely that these numbers are inflated (i.e., likely to be lower when the total global market output is factored in). Gross productive output data is secured from OECD databases.

Table 4.59 Domestic global market ratio: Sweden

Industry	2000	2001	2002	2003	2004	2005	2006	2007	2008	2009	2010
Construction (ISIC Rev.4 Code: 41)	0.83	0.80	0.87	0.95	1.04	0.99	1.03	1.17	1.20	1.10	1.29
Automotive Retail & Repair (ISIC Rev.4 Code: 45)	2.22	2.08	2.13	2.35	2.75	2.62	2.56	2.67	2.56	2.48	2.96
Wholesale Trade (ISIC Rev.4 Code: 46)	2.13	2.04	2.22	2.41	2.37	2.38	2.43	2.53	2.75	2.43	2.64
Retail Trade (ISIC Rev.4 Code: 47)	1.81	1.72	1.84	2.03	2.24	2.26	2.35	2.52	2.54	2.33	2.54
Land Transport (ISIC Rev.4 Code: 49)	1.48	1.37	1.46	1.65	1.69	1.61	1.61	1.76	1.81	1.78	1.83
Water Transport (ISIC Rev.4 Code: 50)	2.60	2.23	2.36	2.66	2.64	2.69	2.64	2.34	2.33	2.29	2.07
Air Transport (ISIC Rev.4 Code: 51)	1.00	1.04	1.13	1.09	1.14	1.11	1.12	1.15	1.15	1.17	0.98
Support Activities for Transport (ISIC Rev.4 Code: 52)											
Postal & Courier Activities (ISIC Rev.4 Code: 53)											
Hotel & Restaurants (ISIC Rev.4 Code: 55, 56)	0.62	0.58	0.61	0.68	0.71	0.69	0.69	0.75	0.80	0.71	0.78
Publishing Activities (ISIC Rev.4 Code: 58)	1.03	0.92	0.92	1.10	1.19	1.17	1.15	1.20	1.22	1.13	1.21
Audiovisual & Broadcasting (ISIC Rev.4 Code: 59, 60)	2.04	1.95	2.16	2.45	2.53	2.49	2.65	2.79	2.89	2.54	2.67
Telecommunications (ISIC Rev.4 Code: 61)	0.45	0.72	0.80	0.93	0.95	0.93	0.96	1.03	1.04	0.94	0.98
IT & Information Services (ISIC Rev.4 Code: 62, 63)	1.61	1.55	1.46	1.72	2.04	2.04	1.89	1.98	2.00	1.69	1.76

Financial Service Activities (ISIC Rev.4 Code: 64)	0.76	0.64	0.59	0.65	0.78	0.75	0.69	0.71	0.70	0.67	0.71
Insurance & Pensions (ISIC Rev.4 Code: 65)	0.54	0.47	0.48	0.51	0.56	0.54	0.53	0.61	0.68	0.54	0.55
Auxiliary Financial Services (ISIC Rev.4 Code: 66)	0.13	0.13	0.13	0.17	0.17	0.17	0.15	0.16	0.18	0.15	0.14
Legal, Accounting & Consultancy (ISIC Rev.4 Code: 69,70)	2.13	1.87	1.91	1.78	1.77	1.73	1.81	1.95	2.01	1.88	2.07
Architectural & Technical Services (ISIC Rev.4 Code: 71)											
Research & Development (ISIC Rev.4 Code: 72)											
Advertising & Market Research (ISIC Rev.4 Code: 73)	3.56	3.36	3.36	3.61	3.63	3.46	3.57	3.75	3.79	3.53	3.83
Other Professionals Services (ISIC Rev.4 Code: 74, 75)	2.14	2.15	2.15	2.43	2.56	2.54	2.78	3.03	3.20	3.10	3.42
Rental & Leasing Activities (ISIC Rev.4 Code: 77)	0.47	0.43	0.49	0.58	0.55	0.52	0.53	0.61	0.65	0.59	0.66
Employment Activities (ISIC Rev.4 Code: 78)	2.12	2.18	2.33	2.51	2.49	2.39	2.32	2.30	2.36	2.17	2.50
Travel Service (ISIC Rev.4 Code: 79)	3.96	3.87	4.08	4.35	4.19	3.97	4.02	4.29	4.10	3.96	4.19
Security & Building Services (ISIC Rev.4 Code: 80, 81, 82)	2.16	2.09	2.04	2.16	2.19	2.09	2.27	2.42	2.47	2.29	2.47
Public Administration & Defense (ISIC Rev.4 Code: 84)	0.81	0.66	0.67	0.77	0.80	0.75	0.77	0.80	0.78	0.66	0.70
Education (ISIC Rev.4 Code: 85)	1.10	1.02	1.08	1.27	1.33	1.28	1.29	1.35	1.37	1.21	1.28

(continued)

Table 4.59 (continued)

Industry	2000	2001	2002	2003	2004	2005	2006	2007	2008	2009	2010
Human Health Activities (ISIC Rev.4 Code: 86)	2.65	2.61	2.77	3.01	3.00	2.87	2.86	2.99	3.00	2.71	2.74
Residential Care & Social Work (ISIC Rev.4 Code: 87, 88)	5.81	5.54	5.77	5.98	5.92	5.58	5.53	5.66	5.65	5.07	5.23
Creative Entertainment Activities (ISIC Rev.4 Code: 90, 91, 92)	2.80	2.65	2.78	3.04	3.07	3.01	3.13	3.24	3.21	2.90	3.06
Sports & Recreation Activities (ISIC Rev.4 Code: 93)	2.53	2.38	2.44	2.67	2.75	2.66	2.71	2.76	2.81	2.62	2.80
Membership Organizations (ISIC Rev.4 Code: 94)	5.15	4.99	5.23	5.63	5.74	5.68	5.66	5.71	5.77	5.22	5.46
Repair Services (ISIC Rev.4 Code: 95)	1.35	1.25	1.34	1.49	1.65	1.67	1.68	1.83	1.87	1.76	1.85
Personal Service Activities (ISIC Rev.4 Code: 96)	1.19	1.17	1.29	1.36	1.46	1.47	1.51	1.66	1.75	1.62	1.82
Undifferentiated Services (ISIC Rev.4 Code: 97, 98)	0.07	0.08	0.09	0.10	0.12	0.13	0.15	0.17	0.17	0.16	0.17
All Services (Average)	0.95	0.87	0.91	1.04	1.09	1.06	1.07	1.15	1.19	1.06	1.12
Business Sector Services (Average)	0.92	0.85	0.89	1.01	1.07	1.04	1.05	1.14	1.19	1.07	1.13
Business Sector Services (Average excluding real estate)	0.90	0.84	0.87	0.99	1.05	1.03	1.04	1.13	1.18	1.06	1.12
All Industries (Average)	0.76	0.72	0.77	0.86	0.89	0.84	0.83	0.86	0.84	0.75	0.75

Note: DGM (Domestic Global Market) Ratio is computed by dividing gross productive output of a country by the OECD productive output in the corresponding service sector. In this computation the OECD output is taken as a proxy for global market output. Therefore, it is likely that these numbers are inflated (i.e., likely to be lower when the total global market output is factored in). Gross productive output data is secured from OECD databases.

Table 4.60 Domestic global market ratio: United States

Industry	2000	2001	2002	2003	2004	2005	2006	2007	2008	2009	2010
Construction (ISIC Rev.4 Code: 41)	35.48	36.69	35.98	34.41	33.72	34.25	33.62	31.15	29.23	28.64	26.59
Automotive Retail & Repair (ISIC Rev.4 Code: 45)											
Wholesale Trade (ISIC Rev.4 Code: 46)											
Retail Trade (ISIC Rev.4 Code: 47)											
Land Transport (ISIC Rev.4 Code: 49)	34.36	34.73	34.39	32.24	31.88	32.27	32.33	30.79	30.21	29.22	29.72
Water Transport (ISIC Rev.4 Code: 50)	21.43	21.00	20.75	19.17	17.63	15.62	14.87	13.77	13.35	16.12	14.34
Air Transport (ISIC Rev.4 Code: 51)	43.61	42.80	41.68	42.35	41.13	40.36	40.00	39.36	39.21	39.85	40.68
Support Activities for Transport (ISIC Rev.4 Code: 52)	30.07	30.39	29.86	28.24	27.13	26.55	26.51	25.38	24.92	25.46	25.77
Postal & Courier Activities (ISIC Rev.4 Code: 53)	41.05	41.69	41.18	39.67	38.26	37.45	37.77	36.52	35.86	36.56	35.72
Hotel & Restaurants (ISIC Rev.4 Code: 55, 56)	42.41	43.12	42.98	41.74	41.30	41.34	41.25	40.34	39.80	40.23	39.71
Publishing Activities (ISIC Rev.4 Code: 58)	47.64	48.42	48.09	46.58	45.91	45.59	45.20	44.46	44.12	44.52	44.96
Audiovisual & Broadcasting (ISIC Rev.4 Code: 59, 60)											
Telecommunications (ISIC Rev.4 Code: 61)	27.94	47.46	46.21	44.12	43.02	42.87	42.91	42.41	42.40	42.83	42.85

(continued)

Table 4.60 (continued)

Industry	2000	2001	2002	2003	2004	2005	2006	2007	2008	2009	2010
IT & Information Services (ISIC Rev.4 Code: 62, 63)	41.93	41.01	40.18	38.50	37.29	37.10	37.13	36.67	36.26	36.71	36.88
Financial Service Activities (ISIC Rev.4 Code: 64)	41.96	44.10	44.79	43.64	42.45	42.91	43.32	42.65	42.41	42.45	41.66
Insurance & Pensions (ISIC Rev.4 Code: 65)	51.41	52.26	51.76	49.98	49.51	49.47	49.34	48.99	48.79	48.93	48.64
Auxiliary Financial Services (ISIC Rev.4 Code: 66)	58.56	57.84	56.69	55.54	54.94	55.43	55.98	55.43	53.99	54.82	55.74
Legal, Accounting & Consultancy (ISIC Rev.4 Code: 69,70)											
Architectural & Technical Services (ISIC Rev.4 Code: 71)											
Research & Development (ISIC Rev.4 Code: 72)											
Advertising & Market Research (ISIC Rev.4 Code: 73)											
Other Professionals Services (ISIC Rev.4 Code: 74, 75)											
Rental & Leasing Activities (ISIC Rev.4 Code: 77)	41.95	41.84	40.96	40.07	39.38	39.60	39.80	38.58	38.72	38.93	38.32
Employment Activities (ISIC Rev.4 Code: 78)											
Travel Service (ISIC Rev.4 Code: 79)											
Security & Building Services (ISIC Rev.4 Code: 80, 81, 82)											
Public Administration & Defense (ISIC Rev.4 Code: 84)	46.32	47.63	47.74	46.48	45.81	45.83	45.84	45.31	45.27	45.89	45.81

Education (ISIC Rev.4 Code: 85)	43.09	44.27	43.57	42.04	41.17	41.00	41.13	40.47	40.42	40.97	40.30
Human Health Activities (ISIC Rev.4 Code: 86)											
Residential Care & Social Work (ISIC Rev.4 Code: 87, 88)											
Creative Entertainment Activities (ISIC Rev.4 Code: 90, 91, 92)											
Sports & Recreation Activities (ISIC Rev.4 Code: 93)											
Membership Organizations (ISIC Rev.4 Code: 94)											
Repair Services (ISIC Rev.4 Code: 95)											
Personal Service Activities (ISIC Rev.4 Code: 96)											
Undifferentiated Services (ISIC Rev.4 Code: 97, 98)	26.32	25.81	24.36	23.50	22.59	21.77	22.24	22.12	21.60	20.39	18.05
All Services (Average)	42.85	43.48	43.15	41.62	40.83	40.81	40.73	39.80	39.21	39.87	39.41
Business Sector Services (Average)	42.32	42.72	42.31	40.75	40.02	40.05	39.96	38.94	38.17	38.79	38.33
Business Sector Services (Average excluding real estate)	42.01	42.23	41.80	40.28	39.51	39.46	39.44	38.46	37.71	38.32	38.06
All Industries (Average)	29.97	31.25	31.41	29.94	28.49	27.86	27.01	25.36	23.79	24.24	22.23

Note: DGM (Domestic Global Market) Ratio is computed by dividing gross productive output of a country by the OECD productive output in the corresponding service sector. In this computation the OECD output is taken as a proxy for global market output. Therefore, it is likely that these numbers are inflated (i.e., likely to be lower when the total global market output is factored in). Gross productive output data is secured from OECD databases.

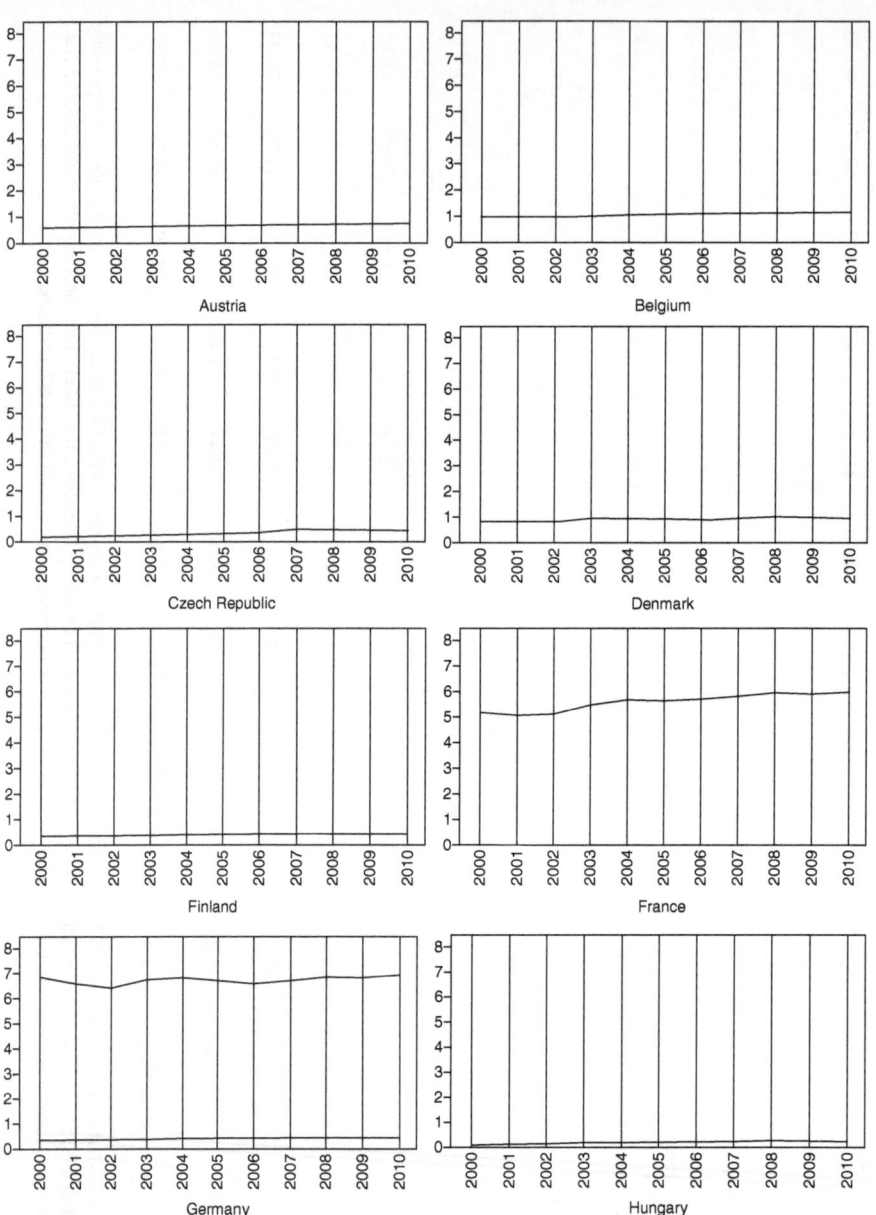

Fig. 4.4 Domestic global market ratio for service sector across countries (2000–2010)

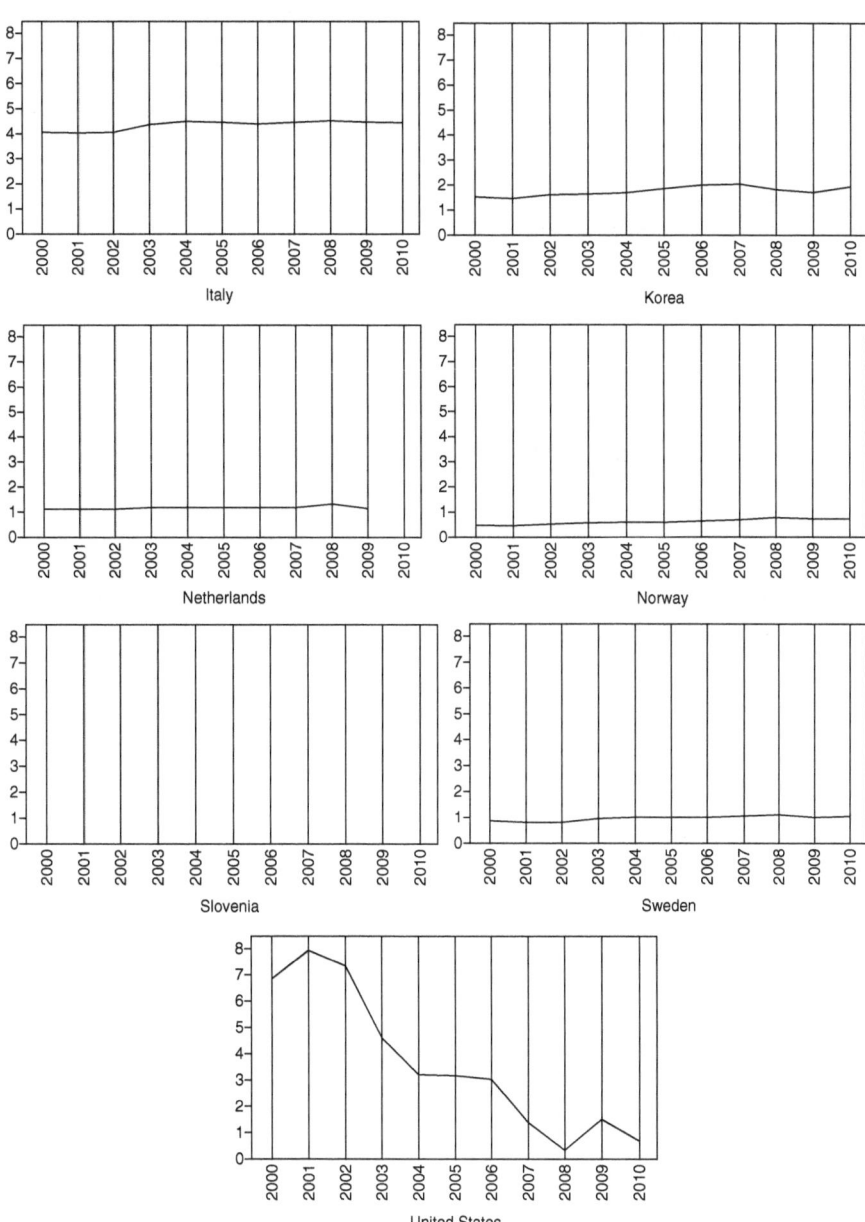

Fig. 4.4 (continued)

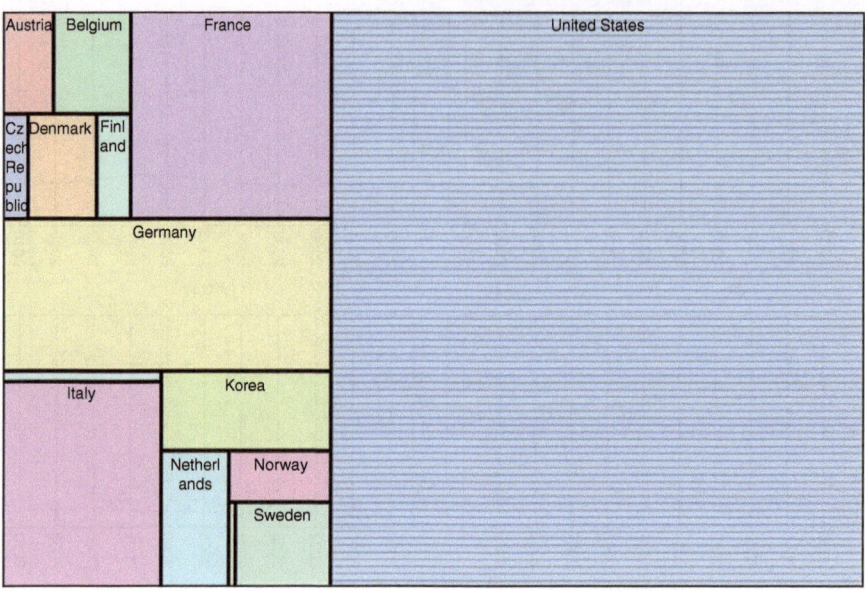

Fig. 4.5 Treemap of countries sized by domestic global market ratio (2000–2010)

Data Source

Organisation for Economic Co-operation and Development (OECD). https://data.oecd.org/

Global Service Sector Data

5

5.1 Gross Profit Ratio (Global Average)

Gross Profit Ratio refers to the ratio of gross profits made by a firm on its sales and is measured by dividing gross profit by sales revenue, while gross profit is measured as the sales revenue minus the cost of goods sold. Gross Profit Ratio is an indicator of the degree of profitability of an industry. At a broad level, industries with relatively higher Gross Profit Ratio would be considered more desirable than industries with lower Gross Profit Ratio. Gross Profit Ratio indicates the amount of funds available to cover other expenses and provide for net profits after paying for the cost of goods. One advantage of using Gross Profit Ratio compared to net profit margin is that the impact of accounting treatments (e.g., depreciation) which could vary across firms and industries is far less, as fixed costs are not factored in its calculation.

Table 5.1 presents the global average of Gross Profit Ratio across 2-Digit SIC industries, while Table 5.2 presents the global average of Gross Profit Ratio across 3-Digit SIC industries. The average Gross Profit Ratio across the entire service sector is presented graphically in Figure 5.1 for 2-Digit SIC industries and Figure 5.2 for 3-Digit SIC industries. This variable is derived from firm level data presented in the GlobalVantage database of Standard & Poor's. The author classified each service firm in the database (about 13,500 firms) into the various industry categories and then computed the average for each of the industries at the 2-Digit and 3-Digit SIC level.

A review of Figures 5.1 and 5.2 indicates that Gross Profit Ratio was significantly impacted in 2009 and 2011, due to the economic shock that happened in 2008, as well as a milder shock in 2010. Incidentally, in both instances profit margins rebounded quite well in the immediate years. During the years 2004–2013, Gross Profit Ratio tended to vary between 39 and 21 % when measured at the 2-Digit SIC level with an average of 34 %, and between 36 and 16 % when measured at the 3-Digit SIC level, with an average of 32 %.

© Springer International Publishing Switzerland 2015
B. Elango, *Service Industry Databook: Understanding and Analyzing Sector Specific Data Across 15 Nations*, DOI 10.1007/978-3-319-19111-9_5

Table 5.1 Gross profit ratio (global average, 2-Digit SIC)

SIC	Industry	2004	2005	2006	2007	2008	2009	2010	2011	2012	2013
40	Railroad Transportation	47.3	30.3	32.6	31.7	30.4	29.8	30.6	29.3	31.5	31.8
41	Local & Interurban Passenger Transit	41.9	32.5	29.1	30.7	30.1	33.1	32.6	29.7	32.1	29.1
42	Trucking & Warehousing	32.3	27.2	25.9	28.7	29.3	28.0	6.5	28.9	28.2	25.4
44	Water Transportation	44.3	37.1	31.2	36.8	36.5	36.3	35.8	34.2	32.2	33.4
45	Transportation by Air	42.7	37.2	33.9	31.0	27.6	28.5	31.1	30.1	28.7	28.8
47	Transportation Services	44.5	40.4	39.5	38.9	38.5	39.4	40.9	38.5	38.8	38.7
48	Communications	46.9	9.8	41.7	46.4	38.7	−37.1	47.3	48.3	47.7	47.3
49	Electric, Gas, & Sanitary Services	49.8	43.8	30.6	40.3	37.8	38.2	39.3	38.6	38.8	39.8
50	Wholesale Trade – Durable Goods	24.8	22.7	21.4	20.0	20.4	2.9	7.3	−1.1	21.1	21.1
51	Wholesale Trade – Nondurable Goods	25.1	19.9	17.6	18.4	20.4	20.9	22.3	21.6	20.9	21.0
52	Building Materials & Supplies	41.7	34.7	35.2	34.0	33.9	34.2	34.0	34.2	33.7	35.0
53	General Merchandise Stores	36.0	33.4	33.3	31.7	32.0	30.8	32.3	32.2	33.2	33.5
54	Food Stores	35.7	31.7	28.7	27.7	28.3	28.9	29.1	31.4	28.8	29.0
55	Automotive Dealers & Service Stations	26.2	24.8	23.3	22.9	22.4	44.6	23.0	22.1	22.4	23.4
56	Apparel & Accessory Stores	48.4	48.2	48.5	46.6	48.4	47.2	48.0	49.2	48.8	48.1
57	Furniture & Homefurnishings Stores	35.2	32.9	34.0	34.3	39.4	27.7	14.6	34.7	35.2	34.7
58	Eating & Drinking Places	53.7	50.6	50.0	46.2	45.9	45.3	41.3	45.0	45.5	46.0
59	Miscellaneous Retail	40.8	37.9	29.4	27.9	36.8	37.4	34.7	34.5	37.9	38.0
60	Depository Institutions	43.0	43.4	43.6	−3.0	39.7	41.7	42.3	41.9	41.7	40.6
61	Nondepository Institutions	66.9	51.5	55.0	62.9	57.2	50.7	54.4	62.5	65.5	63.1
62	Security & Commodity Brokers	61.1	59.1	60.5	64.3	58.3	61.2	58.7	60.8	59.7	60.4

Code	Industry										
63	Insurance Carriers	22.7	21.9	20.6	18.6	17.5	19.2	19.3	19.3	20.3	21.6
64	Insurance Agents, Brokers, & Service	40.9	47.2	57.6	57.5	57.0	53.9	59.3	64.6	62.5	58.5
65	Real Estate	39.8	44.0	39.9	36.1	19.4	19.0	25.8	-22.4	30.6	36.9
67	Holding & Other Investment Offices	61.9	71.1	37.1	64.4	52.4	34.1	67.1	69.6	28.6	57.0
70	Hotels & Other Lodging Places	56.0	-0.3	52.4	53.0	53.7	50.9	51.5	50.9	49.1	51.8
72	Personal Services	41.8	41.5	42.3	42.7	42.4	44.6	46.0	48.2	48.6	46.9
73	Business Services	48.2	43.3		42.2	41.4	42.6	40.8		44.1	42.6
75	Auto Repair, Services, & Parking	50.2	45.0	41.0	40.3	40.7	45.5	46.3	46.0	50.3	47.0
78	Motion Pictures	41.9	-2.6	48.5	29.6	33.4	29.9	36.5	35.9	34.6	34.7
79	Amusement & Recreation Services	-27.4	44.7	47.5	47.3	48.5	46.5	48.7	49.9	51.2	49.5
80	Health Services	-51.3	36.3	39.1	38.4	36.8	36.7	36.7	37.1	36.3	35.8
82	Educational Services	41.9	48.5	46.5	40.8	44.5	45.4	44.9	44.3	45.0	44.2
87	Engineering & Management Services	39.8	34.1	28.4	30.9	30.2	29.1	30.7	29.7	24.2	30.3

Note: Gross Profit Ratio is computed by subtracting cost of goods from sales and dividing by sales, and is expressed as a percentage. Cost of goods and sales data at the firm level are secured from GlobalVantage database of Standard & Poor's

Table 5.2 Gross profit ratio (Global average, 3-Digit SIC)

SIC	Industry	2004	2005	2006	2007	2008	2009	2010	2011	2012	2013
401	Railroads	47.3	30.3	32.6	31.7	30.4	29.8	30.6	29.3	31.5	31.8
421	Trucking and Courier Services	29.6	21.4	19.4	22.0	22.4	22.1	23.4	23.0	22.1	19.5
422	Public Warehousing and Storage	41.1	38.1	37.8	40.9	43.3	42.2	-31.6	40.7	42.6	40.4
441	Deep Sea Foreign Trans. of Freight	45.8	41.3	34.9	37.3	36.9	28.5	31.6	28.8	28.2	27.1
451	Air Transportation, Scheduled	34.4	29.1	24.3	21.4	17.4	18.2	21.9	20.8	19.3	17.8
452	Air Transportation, Nonscheduled	25.7	25.0	28.9	29.3	24.5	30.0	32.2	25.6	27.7	28.1
458	Airports, Flying Fields & Services	67.1	58.3	57.9	52.6	51.3	51.8	51.7	53.6	51.2	51.6
473	Freight Transportation Arrangement	37.0	32.8	27.6	25.4	27.2	27.1	26.8	26.5	25.9	26.9
481	Telephone Communication	61.5	58.1	56.0	55.6	54.5		52.9	53.4	53.0	52.0
483	Radio and Television Broadcasting	49.7	42.4	35.8	44.2	42.8	40.1	41.4	43.0	40.9	38.8
484	Cable and Television Services	-4.9	45.3	35.1	41.0	47.4	46.3	48.7	48.6	53.0	55.2
489	Communications Services	45.1	-83.7	30.5	38.8	15.7	-23.9	44.5	46.4	45.1	45.6
491	Electric Services	51.0	43.2	41.7	39.3	35.7	34.9	36.0	35.0	35.2	35.5
492	Gas Production and Distribution	44.8	40.8	38.8	37.7	35.0	35.7	36.0	34.4	32.8	35.0
493	Combination Utility Services	55.1	47.0	39.3	38.8	34.8	35.8	35.1	32.5	32.2	35.5
494	Water Supply	63.6	59.4	55.7	57.2	56.9	57.1	57.9	56.3	55.7	57.0
495	Sanitary Services	44.1	39.7	38.0	38.0	36.6	35.2	37.4	36.8	36.2	37.7
501	Motor Vehicles, Parts & Supply	26.3	26.9	24.3	24.5	24.4	24.2	26.3	23.2	24.0	20.2
503	Construction Materials Wholesale	25.2	21.7	22.6	18.6	20.1	20.7	20.5	20.9	21.0	21.4
504	P & C Equipment Wholesale	28.6	25.4	23.0	21.4	21.8	22.5	23.3	22.2	23.8	22.7
505	Metals and Minerals Wholesale	19.9	16.4	16.3	16.1	20.8	13.5	14.5	13.2	11.0	12.6
506	Electrical Goods Wholesale	22.7	21.5	20.2	16.6	16.2	19.0	8.3	16.0	19.9	20.6

507	Plumbing & Heating Wholesale	38.2	26.3	25.4	26.0	27.3	28.0	28.1	27.8	27.8	27.0
508	Machinery & Supplies Wholesale	28.1	27.1	26.0	23.7	23.1	26.4	25.6	25.8	25.6	24.0
509	Misc. Durable Goods Wholesale	21.2	18.3	20.2	22.9	19.1				20.2	23.1
511	Paper and Paper Products Wholesale	37.2	19.8	23.0	21.8	21.7	23.1	23.1	21.6	21.9	22.1
512	Drugs & Proprietaries Wholesale	33.4	25.0	24.7	22.1	23.0	23.6	24.4	25.5	25.6	25.1
513	Apparel Wholesale	24.1	18.5	21.5	19.2	24.1	25.1	29.0	29.3	26.6	25.6
514	Groceries Wholesale	24.0	20.6	21.2	19.8	20.7	21.1	22.0	21.0	21.4	20.3
515	Farm Products Wholesale	20.2	12.2	16.2	17.0	17.9	16.7	17.9	16.2	16.6	18.8
516	Chemicals Wholesale	17.3	14.4	14.0	14.4	15.3	15.3	17.1	14.8	13.2	15.6
517	Petroleum Wholesale	23.3	21.5	6.9	15.8	19.0	19.7	21.5	19.3	17.1	18.8
519	Misc. Nondurable Goods Wholesale	31.9	19.0	18.9	17.7	20.4	19.7	20.3	22.9	23.4	19.4
521	Construction Materials Supplies	41.5	35.1	35.7	34.2	34.4	34.8	34.5	34.5	33.8	33.7
531	Department Stores	37.6	35.7	35.5	34.5	35.5	35.2	35.7	35.9	36.1	36.6
539	Misc. General Merchandise Stores	29.6	27.8	27.7	24.4	23.3	19.9	23.1	22.8	25.9	25.8
541	Grocery Stores	35.0	30.5	27.9	27.0	27.0	27.4	27.6	30.3	27.6	27.5
553	Auto and Home Supply Stores	45.6	45.8	41.3	39.9	40.6	40.4	41.6	41.5	43.8	44.2
562	Women's Clothing Stores	49.0	42.5	46.6	47.4	46.8	46.4	47.5	47.4	49.4	49.6
565	Family Clothing Stores	44.7	45.7	46.1	44.0	44.3	42.0	42.9	43.2	41.8	41.8
566	Shoe Stores	49.2	45.0	45.0	42.2	42.7	44.3	44.3	44.0	47.5	45.0
571	Home Furnishings Stores	50.4	44.1	47.0	47.9	47.3	46.9	47.6	48.9	49.7	49.4
573	Radio, Television, & Computer Stores	29.6	27.4	28.8	28.1	40.5	15.4	−13.2	28.9	29.4	27.5
581	Eating and Drinking Places	53.7	50.6	50.0	46.2	45.9	45.3	41.3	45.0	45.5	46.0

(continued)

Table 5.2 (continued)

SIC	Industry	2004	2005	2006	2007	2008	2009	2010	2011	2012	2013
591	Drug Stores and Proprietary Stores	25.2	26.8	25.3	25.2	26.4	26.3	26.7	28.6	27.9	28.5
594	Misc. Shopping Goods Stores	39.5	38.3	37.6	38.2	38.0	38.7	28.0	25.9	40.0	39.6
596	Direct Retailers	49.4	43.1	36.7	9.8	40.0	41.1	41.0	39.8	40.1	39.3
599	Retail Stores	49.1	45.0	6.2	45.4	43.9	43.8	43.7	46.7	47.9	50.5
602	Commercial Banks										
603	Savings Institutions										
609	Functions Related to Banking	43.0	43.4	43.6	-3.0	39.7	41.7	42.3	41.9	41.7	40.6
614	Personal Credit Institutions										
615	Business Credit Institutions	98.9	96.1	93.8	89.0	81.3	78.7	85.8	86.8	90.5	88.8
616	Mortgage Bankers & Brokers	34.8	7.0	16.2	36.9	33.0	22.7	22.9	38.1	40.4	37.4
621	Security Brokers, Dealers										
628	Security & Commodity Services	47.8	44.7	57.5	60.9	59.7	63.4	57.4	58.1	55.5	57.0
631	Life Insurance										
632	Medical Service & Health Insurance	22.7	21.9	20.6	18.6	17.5	19.2	19.3	19.3	20.3	21.6
633	Fire, Marine & Casualty Insurance										
635	Surety Insurance										
641	Insurance Agents & Brokers	40.9	47.2	57.6	57.5	57.0	53.9	59.3	64.6	62.5	58.5
651	Real Estate Operations and Lessors	50.1	56.6	52.8	52.8	37.8	48.4	37.3	17.7	38.5	41.4
655	Subdividers and Developers	37.2	42.0	37.8	29.2	-0.9	12.5	18.4	-13.9	19.3	34.4
672	Investment Offices										
679	Miscellaneous Investing	61.9	71.1	37.1	64.4	52.4	34.1	67.1	69.6	28.6	57.0
701	Hotels and Motels	55.6	-2.3	52.1	52.7	53.3	50.7	51.3	50.7	49.0	51.8

732	Credit Reporting and Collection	71.0	65.1	66.8	67.9	64.7	66.3	63.8	61.9	65.6	52.2
733	Mailing, Reproduction, Stenographic	42.0	35.2	34.7	38.5	38.6	35.4	35.2	34.8	34.8	35.7
734	Services to Buildings	23.9	24.9	25.3	30.0	27.5	26.9	30.0	29.1	29.5	30.0
735	Misc. Equipment Rental and Leasing	57.7	61.7	60.4	60.7	62.4	58.2	58.6	57.6	58.2	55.6
736	Personnel Supply Services	34.6	30.5	31.8	31.8	30.4	28.9	29.7	28.9	28.8	28.7
737	Computer and Data Processing	49.4	44.2	130.6	42.6	41.8	43.4	42.3		45.6	43.9
738	Misc. Business Services	40.8	35.9	34.9	37.9	38.0	38.8	39.1	39.5	38.6	37.0
751	Auto Rentals	59.6	53.4	46.9	47.0	46.4	52.1	50.7	49.3	56.9	52.2
781	Motion Picture Production & Services	39.3	−24.2	50.6	26.6	30.1	24.7	33.8	33.7	33.0	31.4
782	Motion Picture Distribution & Services	38.0	44.2	45.0	31.8	39.5	40.9	40.4	37.8	32.6	37.9
783	Motion Picture Theaters	48.7	39.3	40.7	35.5	39.0	39.3	41.3	40.0	38.2	41.8
794	Commercial Sports	41.8	32.0	40.9	39.4	40.9	36.7	43.6	46.6	49.4	49.6
799	Amusement, Recreation Services	−57.0	47.9	50.1	51.3	51.8	50.6	51.7	52.4	53.4	50.6
805	Nursing and Personal Care Facilities	44.2	25.7	25.2	26.8	29.2	29.4	27.4	24.3	23.8	23.2
806	Hospitals		32.0	40.8	38.9	30.9	33.2	34.2	35.0	35.7	35.9
807	Medical and Dental Laboratories	50.4	48.9	50.9	46.8	48.8	48.0	48.7	49.4	44.2	41.0
808	Home Health Care Services	33.6	36.6	36.0	36.3	37.9	37.6	34.6	34.2	26.9	31.8
809	Misc. Health and Allied Services	44.0	39.2	39.3	41.8	43.6	42.1	41.1	42.5	40.9	40.7

Note: Gross Profit Ratio is computed by subtracting cost of goods from sales and dividing by sales, and is expressed as a percentage. Cost of goods and sales data at the firm level are secured from GlobalVantage database of Standard & Poor's

Fig. 5.1 Gross profit ratio
for the years 2004–2013
(2-Digit SIC)

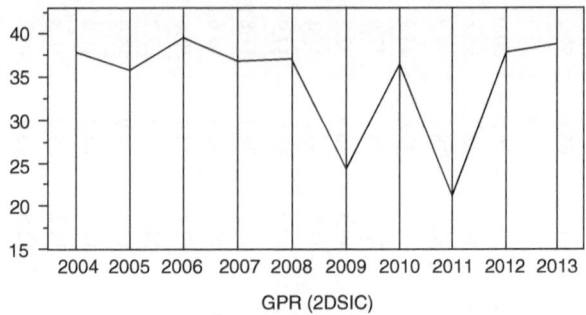

GPR (2DSIC)

Fig. 5.2 Gross profit ratio
for the years 2004–2013
(3-Digit SIC)

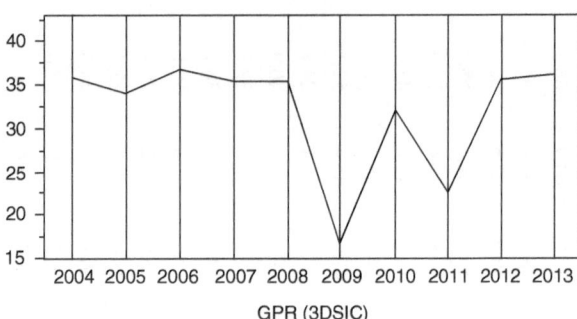

GPR (3DSIC)

5.2   Selling and Administrative Ratio (Global Average)

Selling and Administrative Ratio refers to the extent of expenditure incurred by a
firm for selling and administrative purposes per unit of sales. In many instances this
ratio is also referred to as Selling and General Administrative Ratio. This ratio captures both selling costs (i.e., sales salaries, commission, marketing, advertising
costs, services related to selling) and administrative costs (i.e., finance, human
resources, information technology, legal, and costs related to support, maintenance,
insurance and supplies). While some of the costs in this ratio tend to vary with sales,
a portion of such costs tend to be fixed and thus this ratio serves as a rough proxy
for the overhead and fixed costs incurred by a firm. As noted previously, the inverse
of this ratio has been used as a measure of efficiency of a firm's selling and administrative ability. One disadvantage of this ratio is that it is not possible to determine
which portion of the ratio varies due to sales.

    Table 5.3 presents the global average of the Selling and Administrative Ratio at the
2-Digit SIC level and Table 5.4 presents the same at the 3-Digit SIC level. The average
Selling and Administrative Ratio across services industries is graphically illustrated in
Figure 5.3 at the 2-Digit SIC level, and Figure 5.4 presents the same at the 3-Digit SIC
level. This variable is derived from firm level data presented in the GlobalVantage
database of Standard & Poor's. The author classified each of the service firms in the
database (about 13,500 firms) into various industry categories and computed the average for the each of the industries at the 2-Digit and 3-Digit SIC levels.

Table 5.3 Selling and administrative ratio (Global average, 2-Digit SIC)

SIC	Industry	2004	2005	2006	2007	2008	2009	2010	2011	2012	2013
40	Railroad Transportation	0.10	0.10	0.13	0.13	0.10	0.10	0.10	0.10	0.10	0.15
41	Local & Interurban Passenger Transit	0.19	0.17	0.16	0.13	0.13	0.16	0.17	0.18	0.14	0.14
42	Trucking & Warehousing	0.12	0.12	0.11	0.12	0.12	0.12	0.21	0.12	0.13	0.10
44	Water Transportation	0.12	0.11	0.11	0.10	0.10	0.11	0.17	0.11	0.11	0.11
45	Transportation by Air	0.13	0.13	0.12	0.12	0.11	0.11	0.10	0.11	0.10	0.12
47	Transportation Services	0.24	0.23	0.23	0.22	0.20	1.33	0.29	0.27	0.22	0.22
48	Communications	0.35	0.28	0.29	0.31	0.47	2.47	0.27	0.27	0.25	0.26
49	Electric, Gas, & Sanitary Services	0.16	0.14	0.23	0.20	0.18	0.26	0.35	0.14	0.14	0.13
50	Wholesale Trade – Durable Goods	0.16	0.21	0.21	0.17	0.20	0.18	0.42	0.33	0.24	0.20
51	Wholesale Trade – Nondurable Goods	0.16	0.14	0.24	0.18	0.15	0.15	0.18	0.17	0.21	0.15
52	Building Materials & Supplies	0.25	0.24	0.25	0.25	0.26	0.27	0.26	0.26	0.26	0.32
53	General Merchandise Stores	0.24	0.25	0.25	0.22	0.23	0.32	0.26	0.23	0.23	0.23
54	Food Stores	0.23	0.23	0.23	0.23	0.23	0.23	0.23	0.23	0.23	0.23
55	Automotive Dealers & Service Stations	0.17	0.18	0.16	0.16	0.14	0.46	0.15	0.23	0.15	0.16
56	Apparel & Accessory Stores	0.31	0.32	0.34	0.33	0.44	0.39	0.37	0.38	0.39	0.41
57	Furniture & Homefurnishings Stores	0.25	0.25	0.26	0.27	−0.01	0.31	0.73	0.28	0.29	0.28
58	Eating & Drinking Places	0.37	0.38	0.37	0.37	0.35	0.35	0.35	0.41	0.42	0.38
59	Miscellaneous Retail	0.32	0.37	0.32	0.29	0.35	0.33	0.30	0.29	0.29	0.30
60	Depository Institutions	0.21	0.20	0.20	0.95	0.24	0.20	0.21	0.22	0.23	0.25
61	Nondepository Institutions	0.48	0.39	0.28	0.34	0.33	0.34	0.35	0.34	0.36	0.33
62	Security & Commodity Brokers	0.39	0.36	0.34	0.32	0.34	0.38	0.23	0.28	0.28	0.46
63	Insurance Carriers	0.16	0.16	0.17	0.15	0.16	0.33	0.18	0.17	0.15	0.17
64	Insurance Agents, Brokers, & Service	0.33	0.41	0.33	0.24	0.37	0.39	0.59	0.60	0.45	0.46

(continued)

Table 5.3 (continued)

SIC	Industry	2004	2005	2006	2007	2008	2009	2010	2011	2012	2013
65	Real Estate	0.17	0.22	0.17	0.21	0.24	0.25	0.22	0.23	0.19	0.19
67	Holding & Other Investment Offices	0.95	0.62	1.04	0.54	0.74	0.91	0.36	0.37	1.10	0.47
70	Hotels & Other Lodging Places	0.27	0.26	0.41	0.40	0.36	0.35	0.33	0.40	0.38	0.30
72	Personal Services	0.64	0.28	0.24	0.27	0.30	0.28	0.30	0.36	1.24	0.32
73	Business Services	0.43	0.41	−0.39	0.35	0.87	0.35	0.36	0.37	0.26	0.33
75	Auto Repair, Services, & Parking	0.14	0.15	0.15	0.15	0.14	0.56	0.17	0.18	0.23	0.20
78	Motion Pictures	0.35	1.23	0.05	0.34	0.43	0.33	0.33	0.29	0.31	0.33
79	Amusement & Recreation Services		0.50	0.35	0.32	0.70	0.38	0.42	0.37	0.40	0.40
80	Health Services		0.50	0.51	0.67	0.26	0.24	0.23	0.21	0.21	0.21
82	Educational Services	0.31	0.29	0.31	0.36	0.30	0.31	0.28	0.28	0.29	0.33
87	Engineering & Management Services	0.33	1.34	0.54	0.27	0.24	0.22	0.23	0.23	0.24	0.30

Note: Selling and Administrative Ratio is computed by factoring in all operational costs unrelated directly to production of services and dividing the result by sales. Selling and administrative expenses and sales data at the firm level are secured from GlobalVantage database of Standard & Poor's

Table 5.4 Selling and administrative ratio (Global average, 3-Digit SIC)

SIC	Industry	2004	2005	2006	2007	2008	2009	2010	2011	2012	2013
401	Railroads	0.10	0.10	0.13	0.13	0.10	0.10	0.10	0.10	0.10	0.15
421	Trucking and Courier Services	0.11	0.08	0.08	0.10	0.09	0.10	0.10	0.10	0.10	0.09
422	Public Warehousing and Storage	0.14	0.15	0.19	0.17	0.18	0.17	0.46	0.15	0.16	0.13
441	Deep Sea Foreign Trans. of Freight	0.07	0.08	0.09	0.09	0.08	0.10	0.11	0.10	0.11	0.11
451	Air Transportation, Scheduled	0.13	0.12	0.11	0.10	0.09	0.10	0.10	0.10	0.09	0.11
452	Air Transportation, Nonscheduled	0.12	0.10	0.09	0.12	0.12	0.16	0.13	0.11	0.10	0.16
458	Airports, Flying Fields & Services	0.15	0.17	0.15	0.16	0.16	0.13	0.12	0.14	0.13	0.13
473	Freight Transportation Arrangement	0.20	0.20	0.16	0.19	0.16	0.20	0.23	0.23	0.15	0.17
481	Telephone Communication	0.32	0.24	0.21	0.21	0.23	6.32	0.26	0.25	0.23	0.24
483	Radio and Television Broadcasting	0.39	0.28	0.27	0.26	0.25	0.26	0.25	0.28	0.26	0.24
484	Cable and Television Services	0.54	0.28	0.43	0.36	0.35	0.23	0.22	0.25	0.25	0.23
489	Communications Services	0.30	0.33	0.37	0.46	0.93	0.57	0.31	0.30	0.27	0.28
491	Electric Services	0.16	0.12	0.15	0.11	0.14	0.12	0.11	0.11	0.11	0.10
492	Gas Production and Distribution	0.16	0.13	0.26	0.11	0.11	0.11	0.11	0.11	0.10	0.11
493	Combination Utility Services	0.11	0.11	0.12	0.13	0.11	0.12	0.11	0.13	0.13	0.10
494	Water Supply	0.17	0.15	0.16	0.21	0.19	1.42	2.79	0.25	0.19	0.20
495	Sanitary Services	0.17	0.21	0.18	0.16	0.15	0.16	0.16	0.16	0.16	0.16
501	Motor Vehicles, Parts & Supply	0.25	0.20	0.17	0.16	0.16	0.16	0.49	0.16	0.16	0.32
503	Construction Materials Wholesale	0.13	0.13	0.15	0.14	0.15	0.21	0.16	0.16	0.18	0.46
504	P & C Equipment Wholesale	0.17	0.23	0.31	0.17	0.21	0.18	0.24	0.30	0.52	0.20
505	Metals and Minerals Wholesale	0.22	0.35	0.47	0.22	0.11	0.15	0.11	0.10	0.12	0.10
506	Electrical Goods Wholesale	0.13	0.23	0.15	0.16	0.14	0.18	1.07	0.67	0.22	0.20
507	Plumbing & Heating Wholesale	0.18	0.18	0.16	0.16	0.17	0.20	0.20	0.18	0.19	0.19
508	Machinery & Supplies Wholesale	0.16	0.15	0.15	0.15	0.14	0.18	0.17	0.17	0.16	0.17

(continued)

Table 5.4 (continued)

SIC	Industry	2004	2005	2006	2007	2008	2009	2010	2011	2012	2013
509	Misc. Durable Goods Wholesale	0.14	0.14	0.15	0.23	0.60	0.20	0.20	0.30	0.20	0.23
511	Paper and Paper Products Wholesale	0.18	0.16	0.17	0.17	0.18	0.18	0.17	0.19	0.17	0.17
512	Drugs & Proprietaries Wholesale	0.20	0.19	0.19	0.17	0.16	0.18	0.18	0.18	0.19	0.17
513	Apparel Wholesale	0.20	0.20	0.23	0.22	0.22	0.23	0.20	0.23	0.21	0.23
514	Groceries Wholesale	0.13	0.13	0.14	0.15	0.15	0.15	0.17	0.17	0.25	0.17
515	Farm Products Wholesale	0.37	0.15	0.18	0.33	0.13	0.15	0.16	0.26	0.19	0.14
516	Chemicals Wholesale	0.10	0.10	0.10	0.09	0.09	0.09	0.11	0.08	0.11	0.09
517	Petroleum Wholesale	0.12	0.09	0.49	0.21	0.13	0.08	0.13	0.09	0.09	0.10
519	Misc. Nondurable Goods Wholesale	0.16	0.13	0.19	0.19	0.15	0.15	0.60	0.35	0.83	0.19
521	Construction Materials Supplies	0.26	0.25	0.26	0.26	0.27	0.28	0.27	0.27	0.27	0.27
531	Department Stores	0.24	0.23	0.24	0.23	0.24	0.38	0.24	0.23	0.24	0.24
539	Misc. General Merchandise Stores	0.25	0.28	0.28	0.20	0.19	0.21	0.29	0.20	0.19	0.20
541	Grocery Stores	0.22	0.22	0.21	0.22	0.22	0.22	0.22	0.21	0.21	0.22
553	Auto and Home Supply Stores	0.31	0.31	0.31	0.29	0.30	0.34	0.31	0.30	0.33	0.33
562	Women's Clothing Stores	0.33	0.35	0.34	0.35	0.36	0.39	0.38	0.38	0.41	0.41
565	Family Clothing Stores	0.27	0.29	0.30	0.31	0.32	0.32	0.33	0.33	0.33	0.32
566	Shoe Stores	0.30	0.29	0.32	0.33	0.34	0.36	0.36	0.36	0.38	0.33
571	Home Furnishings Stores	0.32	0.31	0.37	0.38	0.38	0.39	0.38	0.38	0.38	0.39
573	Radio, Television, & Computer Stores	0.22	0.23	0.24	0.25	−0.35	0.31	1.20	0.25	0.26	0.24
581	Eating and Drinking Places	0.37	0.38	0.37	0.37	0.35	0.35	0.35	0.41	0.42	0.38

Code	Industry										
591	Drug Stores and Proprietary Stores	0.20	0.49	0.21	0.20	0.20	0.20	0.20	0.21	0.20	0.20
594	Misc. Shopping Goods Stores	0.28	0.28	0.28	0.28	0.28	0.30	0.30	0.30	0.31	0.31
596	Direct Retailers	0.46	0.48	0.47	0.37	0.55	0.48	0.35	0.32	0.33	0.35
599	Retail Stores	0.35	0.35	0.34	0.34	0.36	0.35	0.36	0.40	0.36	0.39
602	Commercial Banks										
603	Savings Institutions										
609	Functions Related to Banking	0.21	0.20	0.20	0.95	0.24	0.20	0.21	0.22	0.23	0.25
614	Personal Credit Institutions										
615	Business Credit Institutions	0.54	0.58	0.43	0.43	0.47	0.37	0.33	0.36	0.42	0.44
616	Mortgage Bankers & Brokers	0.41	0.20	0.14	0.25	0.19	0.32	0.36	0.32	0.30	0.22
621	Security Brokers, Dealers										
628	Security & Commodity Services	0.28	0.30	0.29	0.27	0.34	0.34	0.20	0.24	0.21	0.48
631	Life Insurance										
632	Medical Service & Health Insurance	0.16	0.16	0.17	0.15	0.16	0.33	0.18	0.17	0.15	0.17
633	Fire, Marine & Casualty Insurance										
635	Surety Insurance										
641	Insurance Agents & Brokers	0.33	0.41	0.33	0.24	0.37	0.39	0.59	0.60	0.45	0.46
651	Real Estate Operations and Lessors	0.21	0.41	0.26	0.29	0.43	0.42	0.27	0.49	0.21	0.17
655	Subdividers and Developers	0.18	0.19	0.15	0.20	0.18	0.22	0.22	0.22	0.20	0.21
672	Investment Offices										
679	Miscellaneous Investing	0.95	0.62	1.04	0.54	0.74	0.91	0.36	0.37	1.10	0.47
701	Hotels and Motels	0.26	0.26	0.42	0.41	0.36	0.35	0.32	0.39	0.38	0.29
732	Credit Reporting and Collection	0.40	0.36	0.36	0.38	0.34	0.35	0.37	0.37	0.37	0.39
733	Mailing, Reproduction, Stenographic	0.23	0.22	0.22	0.22	0.22	0.37	0.28	0.25	0.33	0.30
734	Services to Buildings	0.17	0.21	0.19	0.19	0.20	0.19	0.22	0.20	0.19	0.20
735	Misc. Equipment Rental and Leasing	0.23	0.21	0.29	0.24	0.23	0.22	0.22	0.23	0.22	0.25

(continued)

Table 5.4 (continued)

SIC	Industry	2004	2005	2006	2007	2008	2009	2010	2011	2012	2013
736	Personnel Supply Services	0.28	0.26	0.25	0.24	0.27	0.28	0.26	0.24	0.24	0.23
737	Computer and Data Processing	0.46	0.45	0.59	0.38	1.04	0.37	0.38	0.40	0.25	0.34
738	Misc. Business Services	0.50	0.33	0.35	0.25	0.26	0.33	0.25	0.30	0.29	0.27
751	Auto Rentals	0.13	0.13	0.13	0.14	0.13	0.21	0.18	0.17	0.24	0.22
781	Motion Picture Production & Services	0.38	1.73	−0.04	0.36	0.48	0.35	0.36	0.28	0.32	0.30
782	Motion Picture Distribution & Services	0.29	0.27	0.32	0.43	0.46	0.41	0.35	0.37	0.36	0.53
783	Motion Picture Theaters	0.28	0.25	0.19	0.18	0.21	0.21	0.20	0.22	0.23	0.23
794	Commercial Sports	0.24	0.30	0.31	0.28	0.33	0.37	0.32	0.38	0.45	0.47
799	Amusement, Recreation Services		0.28	0.30	0.34	0.87	0.41	0.47	0.38	0.41	0.41
805	Nursing and Personal Care Facilities	0.14	0.09	0.10	0.09	0.11	0.13	0.09	0.10	0.09	0.09
806	Hospitals		0.47	0.88		0.17	0.17	0.16	0.16	0.15	0.18
807	Medical and Dental Laboratories	0.63	0.71	0.69	0.68	0.77	0.61	0.56	0.39	0.38	0.35
808	Home Health Care Services	0.23	0.26	0.26	0.27	0.28	0.27	0.26	0.27	0.29	0.24
809	Misc. Health and Allied Services	0.45	0.81	0.33	0.39	0.27	0.28	0.27	0.27	0.29	0.29
871	Engineering, Architectural Services	0.18	0.16	0.16	0.13	0.14	0.14	0.14	0.13	0.14	0.26
872	Accounting, Auditing & Bookkeeping	0.28	0.27	0.24	0.25	0.22	0.20	0.23	0.22	0.23	0.23
873	Research and Testing Services					0.53	0.38	0.47	0.48	0.45	0.60
874	Management & Public Relations	0.26	0.28	0.28	0.26	0.25	0.28	0.27	0.28	0.33	0.26

Note: Selling and Administrative Ratio is computed by factoring in all operational costs unrelated directly to production of services and dividing the result by sales. Selling and administrative expenses and sales data at the firm level are secured from GlobalVantage database of Standard & Poor's

Fig. 5.3 Selling and administrative ratio for the years 2004–2013 (2-Digit SIC)

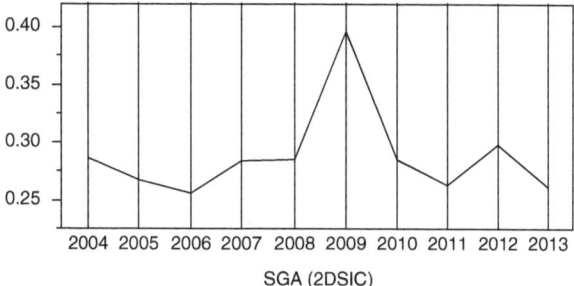

SGA (2DSIC)

Fig. 5.4 Selling and administrative ratio for the years 2004–2013 (3-Digit SIC)

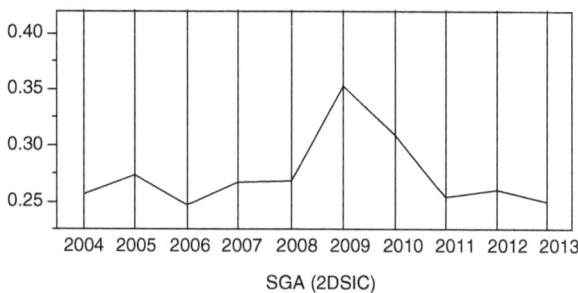

SGA (2DSIC)

A review of Figures 5.3 and 5.4 indicates one major peak for the period for both the 2-Digit SIC level and the 3-Digit SIC level. There is an upward trend in 2012 at the 2-Digit SIC level, a pattern which is not obvious for the 3-Digit level. The 2009 peak can be attributed to a global economic recession in 2008 where all firms faced reduced sales. Selling and Administrative Ratio tends to vary roughly from .39 to 24 with an average of .29 during the years 2004–2013 at the 2-Digit SIC level, which is very comparable to the range of .35–.25 with an average of .28 at the 3-Digit SIC level.

5.3   Leverage (Global Average)

Leverage is measured as the ratio of total liabilities owned by a firm to total assets held by a firm. In other words, this variable shows to what extent a company's total assets are owned by creditors. This variable is also referred to as Financial Leverage or Firm Leverage. In some instances this variable is also measured as debt divided by the equity of the firm, which is not the case here, although the implications of the variable in both operationalizations are the same. Since Leverage represents the extent to which a firm is using borrowed money, it does not represent a higher risk to shareholders of the firm, since creditors need to be paid regularly, irrespective of the financial condition of the firm. However, assuming a firm can earn a higher return than its borrowed cost of capital, it can magnify its returns through leverage rather than use its capital alone.

Table 5.5 presents the global average of Leverage across 2-Digit SIC industries, while Table 5.6 presents the global average of Leverage across 3-Digit SIC industries. The average Leverage across the entire service sector is presented graphically

Table 5.5 Leverage (Global average, 2-Digit SIC)

SIC	Industry	2004	2005	2006	2007	2008	2009	2010	2011	2012	2013
40	Railroad Transportation	0.62	0.64	0.61	0.61	0.61	0.60	0.60	0.58	0.58	0.57
41	Local & Interurban Passenger Transit	0.57	0.56	0.57	0.56	0.60	0.60	0.67	0.62	0.64	0.57
42	Trucking & Warehousing	0.50	0.51	0.51	0.50	0.50	0.50	0.52	0.52	0.52	0.52
44	Water Transportation	0.51	0.49	0.50	0.48		0.51	0.51	0.53	0.54	0.54
45	Transportation by Air	0.80	0.62	0.60	0.60	0.63	0.64	0.61	0.63	0.63	0.64
47	Transportation Services	0.54	0.54	0.53	0.54	0.55	0.53	0.51	0.59	0.53	0.52
48	Communications	0.71	0.57	0.55	0.55	0.57	0.61	0.54	0.55	0.59	0.57
49	Electric, Gas, & Sanitary Services	0.56	0.57	0.55	0.56	0.57	0.58	0.58	0.59	0.60	0.59
50	Wholesale Trade – Durable Goods	0.77	0.61	0.67	0.80	1.01	0.63	0.61	0.65	0.90	0.52
51	Wholesale Trade – Nondurable Goods	0.57	0.58	0.56	0.57	0.55	0.54	0.53	0.53	0.54	0.53
52	Building Materials & Supplies	0.59	0.58	0.60	0.58	0.58	0.57	0.55	0.56	0.55	0.53
53	General Merchandise Stores	0.60	0.59	0.59	0.59	0.60	0.60	0.60	0.59	0.57	0.56
54	Food Stores	0.59	0.59	0.58	0.60	0.60	0.59	0.59	0.58	0.57	0.57
55	Automotive Dealers & Service Stations	0.60	0.60	0.60	0.59	0.62	0.64	0.62	0.61	0.61	0.60
56	Apparel & Accessory Stores	0.47	0.48	0.48	0.46	0.58	0.62	0.63	0.63	0.45	0.45
57	Furniture & Homefurnishings Stores	0.55	0.56	0.56	0.56	0.59	0.62	0.59	0.56	0.56	0.55
58	Eating & Drinking Places	0.55	0.54	0.54	0.54	0.54	0.56	0.57	0.57	0.55	0.56
59	Miscellaneous Retail	0.56	0.54	0.52	0.51	0.53	0.54	0.54	0.53	0.52	0.52
60	Depository Institutions	0.90	0.89	0.89	0.89	0.89	0.89	0.89	0.89	0.89	0.89

61	Nondepository Institutions	0.69	0.69	0.70	0.69	0.71	3.09	0.72	0.71	0.80	0.78
62	Security & Commodity Brokers	0.51	0.50	0.51	0.53	0.52	0.55	0.52	0.50	0.51	0.51
63	Insurance Carriers	0.71	0.70	0.69	0.67	0.69	0.69	0.70	0.72	0.73	0.74
64	Insurance Agents, Brokers, & Service	0.61	0.60	0.59	0.59	0.62	0.61	0.59	0.65	0.69	0.71
65	Real Estate	0.56	0.55	0.54	0.52	0.55	0.58	0.71	0.57	0.54	0.51
67	Holding & Other Investment Offices	0.28	0.27	0.30	0.29	0.31	0.31	0.31	0.31	0.31	0.29
70	Hotels & Other Lodging Places	0.45	0.45	0.42	0.40	0.41	0.41	0.41	0.42	0.42	0.41
72	Personal Services	0.75	0.88	0.61	0.61	0.61	0.61	0.60	0.60	0.62	0.58
73	Business Services	0.52	0.50	0.49	0.46	0.48	0.49	0.51	0.48	0.50	0.48
75	Auto Repair, Services, & Parking	0.64	0.64	0.61	0.63	0.66	0.65	0.63	0.63	0.63	0.62
78	Motion Pictures	0.58	0.54	0.52	0.50	0.48	0.47	0.49	0.47	0.47	0.46
79	Amusement & Recreation Services	0.55	0.53	0.55	0.56	0.57	0.56	0.54	0.57	0.56	0.56
80	Health Services	0.51	0.49	0.50	0.50	0.52	0.49	0.51	0.51	0.54	0.52
82	Educational Services				0.40	0.41	0.45	0.46	0.45	0.43	0.43
87	Engineering & Management Services	0.52	0.53	0.53	0.51	0.52	0.53	0.50	0.51	0.52	0.65

Note: Leverage is computed as total liabilities divided by total assets. Total liabilities and total assets at the firm level are secured from Global Vantage database of Standard & Poor's

Table 5.6 Leverage (Global average, 3-Digit SIC)

SIC	Industry	2004	2005	2006	2007	2008	2009	2010	2011	2012	2013
401	Railroads	0.62	0.64	0.61	0.61	0.61	0.60	0.60	0.58	0.58	0.57
421	Trucking and Courier Services	0.54	0.52	0.53	0.53	0.54	0.54	0.56	0.55	0.55	0.56
422	Public Warehousing and Storage	0.46	0.47	0.46	0.44	0.44	0.44	0.48	0.49	0.47	0.48
441	Deep Sea Foreign Trans. of Freight	0.57	0.52	0.51	0.51	0.53	0.55	0.55	0.58	0.59	0.60
451	Air Transportation, Scheduled		0.72	0.69	0.69	0.72	0.74	0.70	0.71	0.72	0.73
452	Air Transportation, Nonscheduled	0.62	0.61	0.58	0.63	0.67	0.62	0.64	0.61	0.65	0.62
458	Airports, Flying Fields & Services	0.38	0.39	0.39	0.39	0.41	0.40	0.39	0.42	0.43	0.44
473	Freight Transportation Arrangement	0.58	0.55	0.53	0.53	0.53	0.53	0.50	0.50	0.51	0.51
481	Telephone Communication	0.97	0.64	0.59	0.57	0.58	0.57	0.56	0.58	0.67	0.63
483	Radio and Television Broadcasting	0.53	0.50	0.51	0.49	0.56	0.58	0.55	0.53	0.54	0.51
484	Cable and Television Services	0.71	0.57	0.62	0.64	0.68	1.21	0.68	0.64	0.63	0.66
489	Communications Services	0.52	0.53	0.52	0.54	0.52	0.51	0.49	0.50	0.52	0.53
491	Electric Services	0.58	0.59	0.57	0.57	0.59	0.59	0.60	0.61	0.64	0.63
492	Gas Production and Distribution	0.57	0.59	0.57	0.57	0.58	0.59	0.57	0.62	0.60	0.57
493	Combination Utility Services	0.64	0.64	0.61	0.61	0.63	0.63	0.63	0.64	0.64	0.62
494	Water Supply	0.50	0.52	0.53	0.53	0.56	0.56	0.57	0.57	0.56	0.55
495	Sanitary Services	0.51	0.50	0.50	0.50	0.52	0.51	0.49	0.48	0.48	0.47
501	Motor Vehicles, Parts & Supply	0.54	0.54	0.49	0.49	0.50	0.49	0.48	0.52	0.47	0.45
503	Construction Materials Wholesale	0.57	0.56	0.58	0.60	0.58	0.56	0.58	0.58	0.59	0.57
504	P & C Equipment Wholesale	0.58	0.58	1.00	0.92	0.56	0.56	0.54	0.54	0.55	0.56
505	Metals and Minerals Wholesale	0.61	0.63	0.69	2.08	0.79	0.74	0.91	0.60	0.70	0.58
506	Electrical Goods Wholesale	0.59	0.57	0.60	0.58	2.25	0.83	0.69	0.98	1.91	0.49
507	Plumbing & Heating Wholesale	0.57	0.57	0.57	0.58	0.58	0.55	0.52	0.52	0.53	0.50
508	Machinery & Supplies Wholesale	0.58	0.57	0.56	0.54	0.54	0.51	0.50	0.50	0.52	0.48
509	Misc. Durable Goods Wholesale			0.51	0.56	0.52	0.50	0.55	0.51	0.53	0.54

511	Paper and Paper Products Wholesale	0.62	0.60	0.58	0.58	0.59	0.57	0.57	0.56	0.56	0.58
512	Drugs & Proprietaries Wholesale	0.65	0.70	0.66	0.65	0.63	0.62	0.59	0.57	0.56	0.55
513	Apparel Wholesale	0.49	0.48	0.50	0.47	0.50	0.47	0.46	0.46	0.48	0.47
514	Groceries Wholesale	0.56	0.54	0.53	0.53	0.50	0.51	0.50	0.50	0.51	0.51
515	Farm Products Wholesale	0.65	0.66	0.60	0.61	0.53	0.55	0.53	0.55	0.55	0.56
516	Chemicals Wholesale	0.51	0.53	0.57	0.65	0.62	0.52	0.52	0.51	0.49	0.47
517	Petroleum Wholesale	0.57	0.57	0.55	0.53	0.52	0.55	0.54	0.56	0.58	0.56
519	Misc. Nondurable Goods Wholesale	0.67	0.69	0.61	0.61	0.63	0.62	0.59	0.59	0.70	0.66
521	Construction Materials Supplies	0.60	0.61	0.64	0.61	0.59	0.59	0.57	0.57	0.57	0.54
531	Department Stores	0.59	0.59	0.58	0.58	0.59	0.60	0.59	0.57	0.56	0.55
539	Misc. General Merchandise Stores	0.64	0.61	0.64	0.64	0.64	0.64	0.65	0.65	0.60	0.59
541	Grocery Stores	0.60	0.61	0.59	0.61	0.61	0.61	0.60	0.58	0.58	0.58
553	Auto and Home Supply Stores	0.58	0.54	0.53	0.54	0.57	0.57	0.56	0.59	0.60	0.58
562	Women's Clothing Stores	0.43	0.44	0.41	0.41	0.47	0.49	0.47	0.54	0.51	0.52
565	Family Clothing Stores	0.46	0.47	0.48	0.46	0.49	0.49	0.47	0.48	0.51	0.49
566	Shoe Stores	0.61	0.58	0.52	0.50	0.44	0.47	0.46	0.44	0.47	0.37
571	Home Furnishings Stores	0.43	0.40	0.40	0.41	0.44	0.43	0.44	0.42	0.40	0.39
573	Radio, Television, & Computer Stores	0.57	0.59	0.59	0.59	0.63	0.69	0.64	0.58	0.58	0.57
581	Eating and Drinking Places	0.55	0.54	0.54	0.54	0.54	0.56	0.57	0.57	0.55	0.56
591	Drug Stores and Proprietary Stores	0.63	0.66	0.64	0.58	0.57	0.56	0.58	0.57	0.56	0.55
594	Misc. Shopping Goods Stores	0.52	0.52	0.51	0.54	0.54	0.55	0.54	0.54	0.53	0.52
596	Direct Retailers	0.63	0.53	0.49	0.48	0.53	0.55	0.53	0.51	0.52	0.54
599	Retail Stores	0.43	0.44	0.43	0.45	0.47	0.48	0.52	0.49	0.47	0.43
602	Commercial Banks	0.90	0.90	0.90	0.90	0.90	0.90	0.90	0.90	0.90	0.90
603	Savings Institutions	0.89	0.90	0.90	0.89	0.89	0.89	0.89	0.90	0.88	0.87

(continued)

Table 5.6 (continued)

SIC	Industry	2004	2005	2006	2007	2008	2009	2010	2011	2012	2013
609	Functions Related to Banking	0.91	0.65	0.66	0.73	0.66	0.70	0.67	0.65	0.66	0.70
614	Personal Credit Institutions	0.70	0.69	0.71	0.71	0.72	0.69	0.68	0.71	0.71	0.70
615	Business Credit Institutions	0.65	0.66	0.65	0.65	0.67	0.67	0.69	0.67	0.91	0.85
616	Mortgage Bankers & Brokers	0.76	0.80	0.81	0.78	0.84	22.72	0.94	0.77	0.78	0.86
621	Security Brokers, Dealers	0.56	0.56	0.58	0.58	0.56	0.57	0.58	0.55	0.56	0.58
628	Security & Commodity Services	0.42	0.41	0.42	0.47	0.45	0.44	0.43	0.40	0.41	0.41
631	Life Insurance	0.87	0.89	0.89	0.87	0.88	0.88	0.88	0.88	0.88	0.88
632	Medical Service & Health Insurance	0.72	0.70	0.71	0.73	0.78	0.70	0.71	0.73	0.88	0.68
633	Fire, Marine & Casualty Insurance	0.71	0.70	0.69	0.69	0.72	0.70	0.72	0.75	0.75	0.74
635	Surety Insurance	0.69	0.68	0.62	0.70	0.77	0.79	0.79	0.82	0.75	0.74
641	Insurance Agents & Brokers	0.61	0.60	0.59	0.59	0.62	0.61	0.59	0.65	0.69	0.71
651	Real Estate Operations and Lessors	0.56	0.55	0.51	0.50	0.54	0.54	0.52	0.52	0.51	0.50
655	Subdividers and Developers	0.56	0.57	0.54	0.50	0.51	0.52	1.03	0.61	0.51	0.51
672	Investment Offices	0.13	0.12	0.12	0.13	0.15	0.15	0.15	0.15	0.14	0.11
679	Miscellaneous Investing	0.36	0.34	0.38	0.36	0.39	0.38	0.38	0.38	0.39	0.37
701	Hotels and Motels	0.44	0.44	0.42	0.40	0.40	0.40	0.41	0.42	0.41	0.41
732	Credit Reporting and Collection	0.44	0.40	0.45	0.59	0.58	0.53	0.50	0.53	0.46	0.46
733	Mailing, Reproduction, Stenographic	0.51	0.55	0.53	0.54	0.57	0.51	0.45	0.51	0.59	0.57
734	Services to Buildings	0.55	0.57	0.57	0.49	0.53	0.55	0.50	0.49	0.51	0.54
735	Misc. Equipment Rental and Leasing	0.63	0.60	0.59	0.59	0.61	0.60	0.60	0.59	0.60	0.59
736	Personnel Supply Services	0.60	0.60	0.54	0.50	0.50	0.52	0.55	0.53	0.56	0.54
737	Computer and Data Processing	0.52	0.49	0.48	0.45	0.47	0.47	0.48	0.47	0.49	0.47
738	Misc. Business Services	0.48	0.49	0.50	0.50	0.52	0.51	0.53	0.54	0.52	0.51
751	Auto Rentals	0.70	0.71	0.71	0.72	0.75	0.74	0.72	0.73	0.70	0.68

Code	Industry										
781	Motion Picture Production & Services	0.55	0.55	0.55	0.51	0.47	0.46	0.49	0.47	0.47	0.45
782	Motion Picture Distribution & Services	0.45	0.44	0.39	0.40	0.41	0.39	0.39	0.38	0.39	0.41
783	Motion Picture Theaters	0.75	0.55	0.50	0.51	0.54	0.51	0.50	0.50	0.51	0.55
794	Commercial Sports	0.54	0.51	0.56	0.62	0.64	0.65	0.59	0.70	0.70	0.71
799	Amusement, Recreation Services	0.56	0.53	0.56	0.55	0.56	0.54	0.53	0.54	0.53	0.54
805	Nursing and Personal Care Facilities	0.77	0.69	0.72	0.68	0.68	0.69	0.72	0.71	0.72	0.70
806	Hospitals	0.51	0.49	0.51	0.51	0.50	0.49	0.51	0.49	0.52	0.46
807	Medical and Dental Laboratories	0.51	0.43	0.39	0.40	0.61	0.40	0.40	0.43	0.49	0.53
808	Home Health Care Services	0.53	0.49	0.49	0.48	0.53	0.51	0.53	0.57	0.63	0.55
809	Misc. Health and Allied Services	0.43	0.46	0.48	0.48	0.48	0.47	0.49	0.51	0.54	0.51
871	Engineering, Architectural Services	0.59	0.60	0.59	0.58	0.58	0.60	0.55	0.54	0.56	0.87
872	Accounting, Auditing & Bookkeeping	0.39	0.50	0.46	0.40	0.42	0.38	0.42	0.44	0.43	0.41
873	Research and Testing Services	0.37	0.36	0.37	0.36	0.39	0.38	0.34	0.36	0.37	0.37
874	Management & Public Relations	0.50	0.51	0.51	0.48	0.51	0.51	0.52	0.51	0.51	0.50

Note: Leverage is computed as total liabilities divided by total assets. Total liabilities and total assets at the firm level are secured from GlobalVantage database of Standard & Poor's

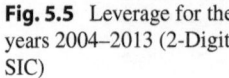

Fig. 5.5 Leverage for the years 2004–2013 (2-Digit SIC)

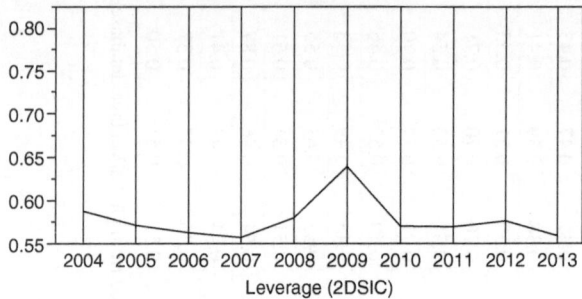

Fig. 5.6 Leverage for the years 2004–2013 (3-Digit SIC)

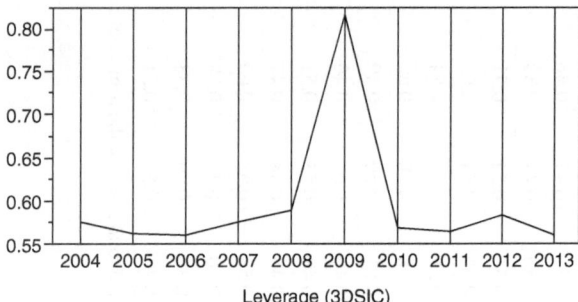

in Figure 5.5 for 2-Digit SIC industries and Figure 5.6 for 3-Digit SIC industries. This variable is derived from firm level data presented in the GlobalVantage database of Standard & Poor's. The author divided total liabilities and total assets for each of the service firms in the database (about 13,500 firms). Then each firm was classified into various industry categories and the author computed the average for the each of the industries at the 2-Digit and 3-Digit SIC levels.

A review of Figures 5.5 and 5.6 indicates that Leverage had a major increase in the year 2009, and this increase is very obvious on the graph at the 3-Digit SIC level. Similar patterns of increase in 2009 were also evident in the case of Gross Profit Ratio and Selling and Administrative Costs. As noted earlier, it is likely that the economic shock of the recession of 2008 is the reason for the peaking of Leverage. Despite the relatively high peak in 2009 as shown in the 3-Digit SIC graph, compared to the 2-Digit SIC graph, the averages across the years 2004–2013 are quite comparable. When measured at the 2-digit SIC level, the average is .57, varying between .64 and .58, whereas when measured at the 3-digit SIC level, the average is .6 and varies between .82 and .56. The latter higher variation in values is expected, due to the fact that industries are more narrowly defined at the 3-Digit level.

5.4 Concentration Ratios (Global)

Concentration refers to the size distribution of firms in an industry (Curry and George 1983). In this chapter, concentration is measured using C4 at both 2-Digit and 3-Digit SIC levels, while C8 is measured at the 2-Digit level alone. C4 represents the market share controlled by the top 4 firms in an industry and C8 represents the market share controlled by the top 8 firms in the industry. The measures indicate the extent to which the largest (four or eight) firms dominate an industry. Concentrated industries could potentially allow for greater stability in firm operations as well as leverage over buyers and suppliers, coupled with higher barriers to entry into the industry (Porter 1980). These conditions could allow some firms to control the market and potentially be more profitable and thus the Concentration Ratio is of interest to managers, investors and public policymakers. While this ratio is typically measured using entire industry sales, the calculation used in this book includes only public firms due to issues related to data availability. Therefore the potential for resulting values to be inflated (i.e., show higher values) should be acknowledged. Due to this limitation, in this book C4 and C8 are represented as CP4 and CP8.

Table 5.7 presents the Global Concentration Ratio (CP4) for the top 4 public companies at the 2-Digit level, Table 5.8 presents the Global Concentration Ratio (CP8) for the top 8 public companies at the 2-Digit level, and Table 5.9 presents the Global Concentration Ratio (CP4) for the top 4 public companies at the 3-Digit level. The average Concentration Ratios are graphically illustrated in Figure 5.7 for CP4 measured at the 2-Digit SIC level, in Figure 5.8 for CP8 measured at the 2-Digit SIC level, and in Figure 5.9 for CP4 when measured at the 3-Digit SIC level. The following process was employed by the author to compute concentration ratios. First, sales revenue for all service firms were downloaded from the GlobalVantage database of Standard & Poor's. These were then grouped by industry and sales were totaled. Third, firms were sorted based on their sales within their industry. Fourth, to compute CP4 the top 4 firms' combined sales were used as the numerator and total industry sales was used as the denominator. In the case of CP8, the top 8 firms were included in the computation.

A review of Figures 5.7, 5.8, and 5.9 shows a similar pattern across the three measures. With the exception of 2007 and 2013, there is an overall pattern of declining concentration over the period 2004–2013. For the years 2007 both C4 and C8 measured at the 2-Digit level show marginal increases of 1.62 and .73 % of concentration. In 2013 there is an increase in concentration of 1.52 % (CP4 2-Digit SIC), 1.63 % (CP8 2-Digit SIC) and .95 % (CP4 3-Digit SIC). While some readers may be concerned with the marked difference in the concentration values across the measures presented, this should not be the case, as each operationalization required the use of a different denominator or numerator.

Table 5.7 Global concentration of top 4 public companies (CP4, 2-Digit SIC)

SIC	Industry	2004	2005	2006	2007	2008	2009	2010	2011	2012	2013
40	Railroad Transportation	0.56	0.53	0.56	0.56	0.56	0.56	0.54	0.54	0.53	0.54
41	Local & Interurban Passenger Transit	0.62	0.61	0.60	0.59	0.59	0.56	0.56	0.56	0.58	0.59
42	Trucking & Warehousing	0.61	0.59	0.62	0.62	0.59	0.56	0.57	0.55	0.54	0.54
44	Water Transportation	0.33	0.34	0.36	0.34	0.34	0.37	0.32	0.33	0.32	0.30
45	Transportation by Air	0.29	0.28	0.27	0.27	0.26	0.27	0.26	0.26	0.24	0.25
47	Transportation Services	0.41	0.40	0.37	0.44	0.34	0.34	0.30	0.29	0.27	0.26
48	Communications	0.22	0.21	0.20	0.21	0.27	0.21	0.20	0.19	0.19	0.19
49	Electric, Gas, & Sanitary Services	0.15	0.14	0.14	0.14	0.16	0.17	0.16	0.15	0.15	0.15
50	Wholesale Trade – Durable Goods	0.40	0.40	0.38	0.36	0.38	0.40	0.34	0.34	0.35	0.33
51	Wholesale Trade – Nondurable Goods	0.37	0.37	0.37	0.33	0.31	0.32	0.32	0.29	0.29	0.33
52	Building Materials & Supplies	0.82	0.82	0.83	0.83	0.80	0.78	0.77	0.75	0.74	0.77
53	General Merchandise Stores	0.51	0.51	0.50	0.51	0.50	0.50	0.48	0.47	0.47	0.48
54	Food Stores	0.33	0.30	0.29	0.60	0.25	0.26	0.25	0.25	0.25	0.27
55	Automotive Dealers & Service Stations	0.34	0.32	0.31	0.29	0.28	0.25	0.25	0.25	0.24	0.23
56	Apparel & Accessory Stores	0.35	0.34	0.32	0.32	0.33	0.33	0.33	0.33	0.32	0.34
57	Furniture & Homefurnishings Stores	0.49	0.48	0.47	0.47	0.45	0.46	0.45	0.42	0.43	0.42
58	Eating & Drinking Places	0.42	0.41	0.39	0.38	0.36	0.35	0.35	0.35	0.35	0.40
59	Miscellaneous Retail	0.37	0.38	0.38	0.42	0.42	0.45	0.46	0.46	0.51	0.53
60	Depository Institutions	0.13	0.13	0.14	0.14	0.12	0.13	0.12	0.11	0.11	0.12

61	Nondepository Institutions	0.59	0.59	0.61	0.69	0.54	0.54	0.68	0.65	0.63	0.63
62	Security & Commodity Brokers	0.50	0.53	0.54	0.53	0.46	0.40	0.35	0.34	0.32	0.34
63	Insurance Carriers	0.20	0.19	0.19	0.33	0.23	0.19	0.18	0.17	0.15	0.16
64	Insurance Agents, Brokers, & Service	0.85	0.84	0.82	0.79	0.79	0.79	0.79	0.79	0.78	0.77
65	Real Estate	0.18	0.14	0.13	0.11	0.61	0.14	0.11	0.11	0.11	0.12
67	Holding & Other Investment Offices	0.29	0.23	0.22	0.22	0.42	0.38	0.23	0.28	0.25	0.23
70	Hotels & Other Lodging Places	0.52	0.54	0.54	0.48	0.44	0.46	0.41	0.39	0.38	0.39
72	Personal Services	0.73	0.72	0.69	0.67	0.67	0.66	0.62	0.60	0.59	0.57
73	Business Services	0.28	0.26	0.23	0.22	0.21	0.22	0.22	0.19	0.19	0.19
75	Auto Repair, Services, & Parking	0.86	0.85	0.78	0.75	0.73	0.69	0.67	0.67	0.68	0.70
78	Motion Pictures	0.58	0.53	0.51	0.47	0.43	0.37	0.33	0.34	0.34	0.35
79	Amusement & Recreation Services	0.48	0.49	0.29	0.28	0.25	0.23	0.21	0.21	0.23	0.24
80	Health Services	0.51	0.49	0.47	0.45	0.45	0.44	0.44	0.42	0.42	0.43
82	Educational Services	0.54	0.54	0.53	0.51	0.47	0.47	0.45	0.44	0.44	0.42
87	Engineering & Management Services	0.34	0.34	0.35	0.34	0.34	0.37	0.39	0.37	0.37	0.40

Note: Global concentration of top 4 public companies is computed by dividing sales revenue of the four largest firms by the sales revenue of all the firms in the industry. Since this variable is computed based on a firm's primary industry of operation, these numbers should be used with caution. Sales at the firm level are secured from GlobalVantage database of Standard & Poor's

Table 5.8 Global concentration of top 8 public companies (CP8, 2-Digit SIC)

SIC	Industry	2004	2005	2006	2007	2008	2009	2010	2011	2012	2013
40	Railroad Transportation	0.80	0.77	0.80	0.78	0.78	0.77	0.76	0.76	0.76	0.76
41	Local & Interurban Passenger Transit	0.81	0.80	0.79	0.77	0.78	0.77	0.77	0.77	0.77	0.80
42	Trucking & Warehousing	0.73	0.72	0.74	0.74	0.71	0.68	0.67	0.65	0.64	0.64
44	Water Transportation	0.48	0.48	0.49	0.48	0.48	0.50	0.45	0.46	0.45	0.44
45	Transportation by Air	0.51	0.47	0.46	0.44	0.42	0.44	0.41	0.41	0.40	0.42
47	Transportation Services	0.61	0.58	0.56	0.61	0.53	0.52	0.48	0.47	0.45	0.44
48	Communications	0.36	0.35	0.34	0.34	0.41	0.35	0.33	0.31	0.31	0.31
49	Electric, Gas, & Sanitary Services	0.24	0.22	0.22	0.22	0.24	0.25	0.24	0.23	0.23	0.23
50	Wholesale Trade – Durable Goods	0.57	0.55	0.53	0.51	0.54	0.57	0.49	0.49	0.50	0.46
51	Wholesale Trade – Nondurable Goods	0.53	0.52	0.51	0.48	0.46	0.46	0.46	0.42	0.43	0.45
52	Building Materials & Supplies	0.42	0.40	0.41	0.43	0.45	0.45	0.48	0.47	0.46	0.49
53	General Merchandise Stores	0.69	0.68	0.68	0.68	0.67	0.67	0.65	0.64	0.63	0.63
54	Food Stores	0.52	0.48	0.46	0.70	0.45	0.45	0.43	0.42	0.42	0.43
55	Automotive Dealers & Service Stations	0.50	0.47	0.47	0.46	0.43	0.41	0.40	0.39	0.38	0.37
56	Apparel & Accessory Stores	0.52	0.51	0.49	0.50	0.49	0.49	0.50	0.50	0.50	0.52
57	Furniture & Homefurnishings Stores	0.64	0.62	0.61	0.61	0.61	0.62	0.61	0.59	0.60	0.57
58	Eating & Drinking Places	0.57	0.56	0.55	0.55	0.53	0.52	0.51	0.52	0.52	0.53
59	Miscellaneous Retail	0.52	0.51	0.52	0.55	0.54	0.57	0.57	0.56	0.60	0.61
60	Depository Institutions	0.24	0.24	0.25	0.24	0.21	0.21	0.20	0.20	0.20	0.22
61	Nondepository Institutions	0.80	0.78	0.78	0.84	0.70	0.71	0.79	0.79	0.78	0.78

62	Security & Commodity Brokers	0.68	0.73	0.71	0.69	0.60	0.55	0.55	0.53	0.51	0.52
63	Insurance Carriers	0.33	0.34	0.33	0.46	0.35	0.30	0.28	0.27	0.26	0.26
64	Insurance Agents, Brokers, & Service	0.95	0.95	0.93	0.92	0.93	0.93	0.93	0.91	0.91	0.90
65	Real Estate	0.28	0.24	0.22	0.18	0.65	0.22	0.18	0.19	0.19	0.20
67	Holding & Other Investment Offices	0.39	0.31	0.30	0.29	0.55	0.47	0.30	0.35	0.33	0.30
70	Hotels & Other Lodging Places	0.68	0.65	0.65	0.60	0.59	0.57	0.54	0.53	0.52	0.53
72	Personal Services	0.92	0.88	0.86	0.84	0.84	0.83	0.81	0.79	0.79	0.78
73	Business Services	0.37	0.34	0.31	0.30	0.29	0.31	0.31	0.28	0.27	0.27
75	Auto Repair, Services, & Parking	0.96	0.94	0.90	0.89	0.88	0.87	0.87	0.87	0.86	0.87
78	Motion Pictures	0.73	0.67	0.66	0.62	0.59	0.56	0.53	0.52	0.53	0.57
79	Amusement & Recreation Services	0.62	0.61	0.41	0.39	0.37	0.36	0.34	0.35	0.37	0.40
80	Health Services	0.69	0.65	0.64	0.63	0.62	0.61	0.60	0.58	0.57	0.61
82	Educational Services	0.72	0.71	0.70	0.67	0.64	0.65	0.65	0.63	0.61	0.58
87	Engineering & Management Services	0.45	0.45	0.46	0.45	0.46	0.48	0.50	0.48	0.49	0.52

Note: Global concentration of top 8 public companies is computed by dividing sales revenue of the eight largest firms by the sales revenue of all the firms in the industry. Since this variable is computed based on a firm's primary industry of operation, these numbers should be used with caution. Sales at the firm level are secured from GlobalVantage database of Standard & Poor's

Table 5.9 Global concentration of top 4 public companies (CP4, 3-Digit SIC)

SIC	Industry	2004	2005	2006	2007	2008	2009	2010	2011	2012	2013
401	Railroads	0.56	0.53	0.56	0.55	0.55	0.55	0.53	0.53	0.53	0.53
421	Trucking and Courier Services	0.66	0.63	0.66	0.67	0.63	0.61	0.63	0.62	0.60	0.61
422	Public Warehousing and Storage	0.58	0.60	0.61	0.62	0.58	0.58	0.59	0.59	0.60	0.63
441	Deep Sea Foreign Trans. of Freight	0.53	0.55	0.57	0.55	0.56	0.60	0.56	0.59	0.59	0.61
451	Air Transportation, Scheduled	0.31	0.30	0.29	0.29	0.28	0.29	0.28	0.27	0.26	0.27
452	Air Transportation, Nonscheduled	0.83	0.83	0.82	0.82	0.79	0.80	0.73	0.72	0.68	0.66
458	Airports, Flying Fields & Services	0.58	0.54	0.49	0.52	0.49	0.45	0.42	0.41	0.39	0.39
473	Freight Transportation Arrangement	0.53	0.51	0.49	0.61	0.48	0.50	0.45	0.45	0.43	0.40
481	Telephone Communication	0.30	0.28	0.27	0.27	0.35	0.28	0.26	0.25	0.25	0.26
483	Radio and Television Broadcasting	0.49	0.47	0.50	0.49	0.49	0.48	0.42	0.41	0.40	0.42
484	Cable and Television Services	0.56	0.55	0.52	0.53	0.53	0.55	0.55	0.58	0.60	0.59
489	Communications Services	0.53	0.52	0.49	0.45	0.42	0.43	0.48	0.48	0.47	0.48
491	Electric Services	0.26	0.24	0.24	0.23	0.26	0.26	0.25	0.24	0.24	0.25
492	Gas Production and Distribution	0.37	0.30	0.28	0.29	0.30	0.30	0.29	0.29	0.31	0.33
493	Combination Utility Services	0.27	0.23	0.22	0.21	0.21	0.21	0.21	0.21	0.23	0.24
494	Water Supply	0.82	0.77	0.77	0.81	0.79	0.77	0.74	0.73	0.71	0.68
495	Sanitary Services	0.45	0.39	0.36	0.36	0.38	0.38	0.39	0.40	0.44	0.49
501	Motor Vehicles, Parts & Supply	0.87	0.88	0.88	0.89	0.88	0.87	0.84	0.84	0.83	0.82
503	Construction Materials Wholesale	0.55	0.55	0.54	0.54	0.51	0.49	0.44	0.45	0.42	0.45
504	P & C Equipment Wholesale	0.48	0.49	0.48	0.46	0.44	0.43	0.42	0.40	0.40	0.43
505	Metals and Minerals Wholesale	0.41	0.38	0.37	0.35	0.37	0.40	0.38	0.39	0.39	0.44
506	Electrical Goods Wholesale	0.36	0.35	0.37	0.38	0.39	0.38	0.39	0.41	0.40	0.44
507	Plumbing & Heating Wholesale	0.78	0.78	0.79	0.79	0.76	0.74	0.72	0.71	0.70	0.74

508	Machinery & Supplies Wholesale	0.34	0.36	0.35	0.34	0.32	0.31	0.30	0.31	0.31	0.32
509	Misc. Durable Goods Wholesale	0.96	0.96	0.94	0.93	0.94	0.95	0.93	0.92	0.93	0.93
511	Paper and Paper Products Wholesale	0.87	0.87	0.87	0.88	0.87	0.85	0.85	0.85	0.83	0.85
512	Drugs & Proprietaries Wholesale	0.76	0.74	0.76	0.73	0.72	0.70	0.69	0.67	0.68	0.69
513	Apparel Wholesale	0.39	0.33	0.31	0.30	0.31	0.40	0.40	0.42	0.43	0.43
514	Groceries Wholesale	0.60	0.59	0.59	0.58	0.58	0.57	0.58	0.59	0.59	0.51
515	Farm Products Wholesale	0.75	0.79	0.81	0.82	0.85	0.84	0.89	0.90	0.91	0.95
516	Chemicals Wholesale	0.61	0.61	0.61	0.67	0.67	0.67	0.66	0.67	0.66	0.63
517	Petroleum Wholesale	0.34	0.35	0.32	0.34	0.34	0.34	0.35	0.36	0.34	0.40
519	Misc. Nondurable Goods Wholesale	0.77	0.68	0.70	0.67	0.64	0.63	0.65	0.65	0.66	0.70
521	Construction Materials Supplies	0.83	0.84	0.85	0.84	0.80	0.79	0.78	0.77	0.76	0.80
531	Department Stores	0.37	0.31	0.39	0.41	0.38	0.34	0.30	0.29	0.29	0.32
539	Misc. General Merchandise Stores	0.80	0.76	0.77	0.75	0.71	0.69	0.67	0.65	0.62	0.61
541	Grocery Stores	0.35	0.31	0.31	0.62	0.27	0.27	0.26	0.26	0.27	0.29
553	Auto and Home Supply Stores	0.80	0.79	0.79	0.76	0.75	0.76	0.77	0.77	0.78	0.80
562	Women's Clothing Stores	0.67	0.66	0.66	0.66	0.66	0.65	0.65	0.66	0.68	0.69
565	Family Clothing Stores	0.64	0.64	0.62	0.61	0.59	0.58	0.59	0.59	0.59	0.61
566	Shoe Stores	0.81	0.80	0.79	0.80	0.78	0.81	0.77	0.77	0.77	0.78
571	Home Furnishings Stores	0.58	0.55	0.57	0.56	0.59	0.61	0.63	0.65	0.66	0.70
573	Radio, Television, & Computer Stores	0.70	0.71	0.69	0.69	0.66	0.67	0.68	0.65	0.66	0.62
581	Eating and Drinking Places	0.42	0.41	0.39	0.38	0.36	0.35	0.35	0.35	0.35	0.40
591	Drug Stores and Proprietary Stores	0.80	0.79	0.79	0.82	0.81	0.80	0.80	0.78	0.81	0.83
594	Misc. Shopping Goods Stores	0.51	0.49	0.50	0.50	0.49	0.50	0.48	0.46	0.46	0.46
596	Direct Retailers	0.50	0.50	0.51	0.51	0.51	0.56	0.60	0.64	0.66	0.68
599	Retail Stores	0.56	0.59	0.60	0.62	0.61	0.52	0.49	0.48	0.49	0.54

(continued)

Table 5.9 (continued)

SIC	Industry	2004	2005	2006	2007	2008	2009	2010	2011	2012	2013
602	Commercial Banks	0.14	0.14	0.14	0.14	0.13	0.13	0.12	0.12	0.11	0.13
603	Savings Institutions	0.91	0.92	0.91	0.79	0.88	0.83	0.81	0.75	0.76	0.75
609	Functions Related to Banking	0.83	0.81	0.75	0.80	0.81	0.79	0.77	0.78	0.79	0.79
614	Personal Credit Institutions	0.66	0.64	0.67	0.67	0.61	0.54	0.57	0.52	0.55	0.55
615	Business Credit Institutions	0.91	0.88	0.87	0.86	0.85	0.82	0.77	0.74	0.72	0.73
616	Mortgage Bankers & Brokers	0.90	0.92	0.91	0.92	0.91	0.88	0.84	0.82	0.81	0.81
621	Security Brokers, Dealers	0.70	0.72	0.73	0.70	0.60	0.58	0.52	0.49	0.47	0.53
628	Security & Commodity Services	0.72	0.75	0.75	0.74	0.70	0.60	0.58	0.58	0.56	0.56
631	Life Insurance	0.32	0.33	0.34	0.39	0.42	0.33	0.31	0.30	0.28	0.28
632	Medical Service & Health Insurance	0.61	0.65	0.70	0.70	0.70	0.70	0.70	0.70	0.68	0.70
633	Fire, Marine & Casualty Insurance	0.44	0.43	0.43	0.42	0.34	0.39	0.36	0.35	0.34	0.33
635	Surety Insurance	0.91	0.92	0.91	1.02	0.98	0.89	0.94	0.95	0.93	0.95
641	Insurance Agents & Brokers	0.85	0.84	0.82	0.79	0.79	0.79	0.79	0.79	0.78	0.77
65 T1	Real Estate Operations and Lessors	0.52	0.47	0.42	0.38	0.63	0.60	0.47	0.48	0.49	0.47
655	Subdividers and Developers	0.21	0.18	0.17	0.16	0.15	0.17	0.15	0.17	0.19	0.24
672	Investment Offices	0.21	0.23	0.19	0.16	0.16	0.17	0.17	0.21	0.28	0.12
679	Miscellaneous Investing	0.30	0.24	0.24	0.23	0.40	0.38	0.26	0.29	0.26	0.24
701	Hotels and Motels	0.53	0.55	0.55	0.49	0.45	0.47	0.42	0.40	0.39	0.40
732	Credit Reporting and Collection	0.87	0.87	0.87	0.86	0.82	0.81	0.84	0.83	0.91	0.88
733	Mailing, Reproduction, Stenographic	0.89	0.87	0.86	0.85	0.82	0.84	0.86	0.87	0.88	0.89
734	Services to Buildings	0.78	0.76	0.76	0.78	0.77	0.70	0.69	0.68	0.65	0.64

735	Misc. Equipment Rental and Leasing	0.49	0.49	0.43	0.40	0.39	0.40	0.40	0.38	0.39	0.39
736	Personnel Supply Services	0.62	0.60	0.58	0.58	0.58	0.56	0.59	0.59	0.56	0.57
737	Computer and Data Processing	0.42	0.39	0.35	0.33	0.32	0.31	0.31	0.28	0.27	0.27
738	Misc. Business Services	0.44	0.41	0.37	0.36	0.36	0.36	0.36	0.34	0.34	0.35
751	Auto Rentals	0.89	0.89	0.83	0.81	0.79	0.77	0.75	0.75	0.77	0.78
781	Motion Picture Production & Services	0.77	0.71	0.69	0.64	0.59	0.55	0.53	0.52	0.51	0.56
782	Motion Picture Distribution & Services	0.95	0.95	0.95	0.96	0.91	0.92	0.93	0.88	0.93	0.94
783	Motion Picture Theaters	0.79	0.72	0.70	0.67	0.66	0.65	0.63	0.62	0.66	0.69
794	Commercial Sports	0.58	0.56	0.52	0.46	0.42	0.44	0.44	0.41	0.41	0.41
799	Amusement, Recreation Services	0.52	0.53	0.32	0.30	0.29	0.27	0.24	0.24	0.25	0.27
805	Nursing and Personal Care Facilities	0.80	0.76	0.74	0.72	0.70	0.69	0.68	0.69	0.70	0.69
806	Hospitals	0.78	0.74	0.72	0.71	0.71	0.70	0.68	0.65	0.64	0.68
807	Medical and Dental Laboratories	0.95	0.94	0.94	0.93	0.90	0.89	0.89	0.90	0.90	0.92
808	Home Health Care Services	0.88	0.88	0.88	0.87	0.86	0.85	0.86	0.85	0.84	0.83
809	Misc. Health and Allied Services	0.90	0.89	0.91	0.90	0.89	0.90	0.90	0.90	0.90	0.91
871	Engineering, Architectural Services	0.35	0.34	0.36	0.36	0.37	0.40	0.43	0.40	0.40	0.44
872	Accounting, Auditing & Bookkeeping	0.49	0.52	0.49	0.48	0.47	0.48	0.49	0.47	0.46	0.47
873	Research and Testing Services	0.60	0.63	0.60	0.59	0.58	0.57	0.58	0.55	0.55	0.56
874	Management & Public Relations	0.66	0.63	0.62	0.62	0.61	0.60	0.57	0.57	0.56	0.56

Note: Global concentration of top 4 public companies is computed by dividing sales revenue of the eight largest firms by the sales revenue of all the firms in the industry. Since this variable is computed based on a firm's primary industry of operation, these numbers should be used with caution. Sales at the firm level are secured from GlobalVantage database of Standard & Poor's

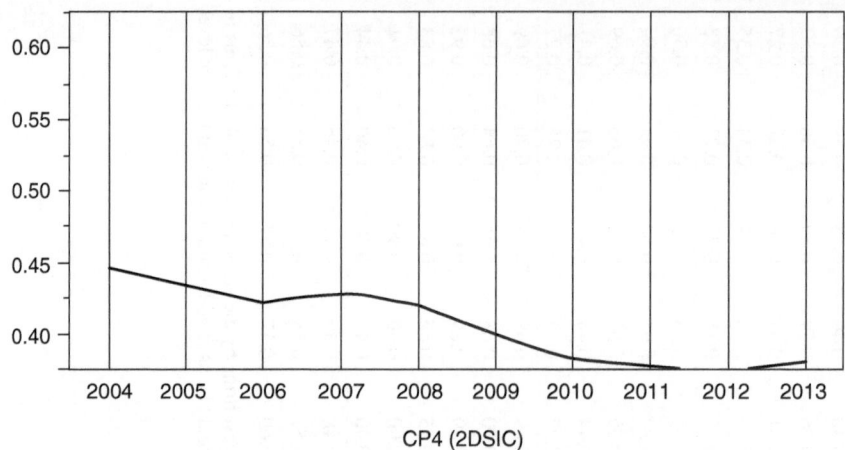

Fig. 5.7 CP4 (concentration ratio) for the years 2004–2013 (2-Digit SIC)

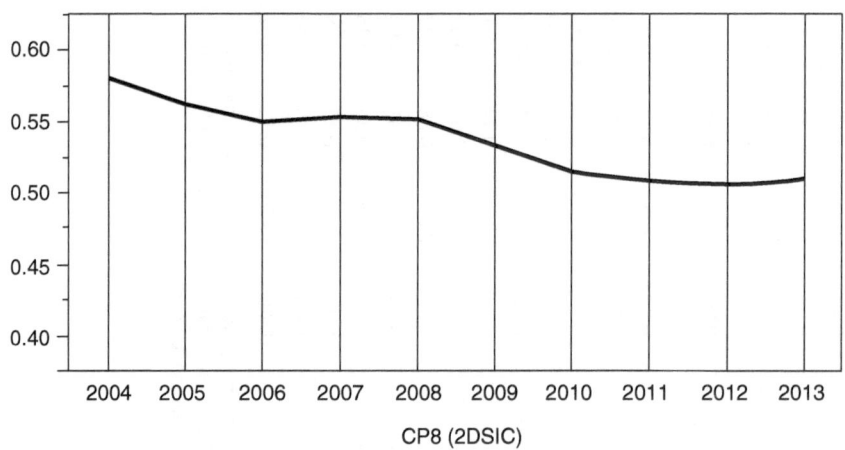

Fig. 5.8 CP8 (concentration ratio) for the years 2004–2013 (2-Digit SIC)

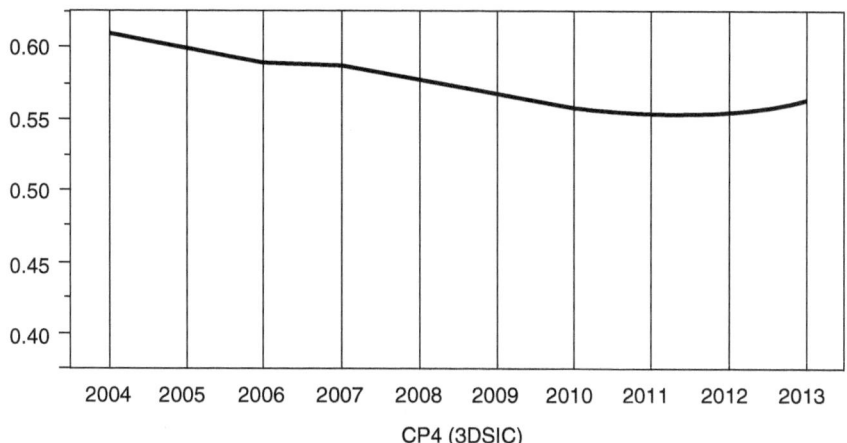

Fig. 5.9 CP4 (concentration ratio) for the years 2004–2013 (3-Digit SIC)

References

Curry B, George KD (1983) Industrial concentration: a survey. J Ind Econ 31(3):203–255
Porter ME (1980) Competitive strategy. Free Press, New York

Data Source

GlobalVantage database (https://www.capitaliq.com/)

Drivers of Profitability in the Service Sector

6

6.1   Logic of the Model

The underlying theme of the model is based on the strategic management paradigm. The field of strategy seeks to understand drivers of firm performance, and the basic premise within the field is that managers can proactively and reactively make decisions which can impact the performance of a firm by factoring in internal strengths/weakness and opportunities/threats in a given environment. While a manager's degree of control to impact performance may vary, the assumption is that, given a particular set of constraints, limitations, or opportunities, firms can undertake particular actions to achieve goals more effectively. The drivers of performance in the strategy management framework can be broadly divided into two sets of factors: factors external to the firm and factors internal to the firm. The conventional guidance offered to managers to achieve firm performance outcomes is to make choices which maximize the fit between the external and internal context. Therefore, each firm's performance will vary based on the effectiveness of its choice of a particular strategy and its implementation, given its own internal resource base and external environmental conditions.[1]

External factors are also referred to as environmental factors, and these are the focus of this chapter. The core idea here is that a firm, however independent and biased towards seeking its goals, is dependent on the environment for its ability to be successful, and therefore outcomes for firm performance can be understood by taking into account the changes in the environment as well as the adaptations made due to environmental influences. External factors can be divided into two groups, namely, macro and industry environment. Typically, macro environment is referred

---

[1] The notions presented in this section are a thematic synthesis of several ideas presented in conventional strategy and organizational theory books. The intent here is not to cover these literatures but to offer a quick rationalization of the model's theoretical underpinnings. Readers interested in learning more may want to refer to books such as: Implanting Strategic Management by H. I. Ansoff and E.J. McDonnell; Organization Theory and Design by R.L. Daft; Contemporary Strategy Analysis by R. M. Grant; Organizational Theory, Design, and Change by G. R. Jones; Readings in The Strategy Process by H. Mintzberg and J. B. Quinn; among others.

© Springer International Publishing Switzerland 2015
B. Elango, *Service Industry Databook: Understanding and Analyzing Sector Specific Data Across 15 Nations*, DOI 10.1007/978-3-319-19111-9_6

to as the broad set of factors within an economy that impact firm outcomes, while industry environment refers to the set of competitors competing with each other. In the empirical models presented in this chapter, relevant external factors for the service industry are captured through incorporation of industry-specific as well as macro environment elements. The notion that external factors impact performance outcomes has been validated in several streams of literature, both conceptually and empirically. For instance, in industrial organization literature, while the emphasis is on industry conditions within the environment, it has been revealed that industry conditions determine firm performance outcomes.[2] Similarly, in organization theory literature, where the emphasis is the macro environment, it has been argued that environmental factors impact performance rather than industry conditions alone.[3]

The goal of this chapter is to decipher empirical patterns between the environmental factors and performance. The intent is to find "useful stylized facts" (Schmalensee 1989: pg. 952) or statistical regularities (Nolle 1991: pg. 60) rather than to make specific causal claims. From the patterns, managers should be able to make reasonable inferences as to the influence of specific elucidations on the impact of these environmental variables on performance, allowing them to make good decisions and thus enhance the chances of performance. The role of environment should not be assumed to have a simple direct impact on firm performance. Extant literature has indicated that these relationships between environment and performance can be direct, mediated, moderated or an interactive influence (Prescott 1986). Therefore, it should not be mistaken that every firm or its managers would be making the same decision in a given environmental context, as the field of strategy subsumes the premise of "equifinality" which essentially accepts the notion that there are differing paths to achieve the same goal. Therefore, different firms would be making differing decisions contingent on their own internal differences, goal-specific needs and extent of freedom offered by the environment (Hrebiniak and Joyce 1985).

6.2   Variables and Relationships

6.2.1   Dependent Variables

The key focus of this chapter's study is to determine the influence of external factors on the performance of service firms, and therefore performance related proxies serve as the dependent variable in all of the regression models. These variables and their operationalizations are derived from extant research and largely borrowed from the prior work of the author. Performance is operationalized using gross profit

---

[2] The logic of this rationalization is referred to as the Structure-Conduct-Performance (S-C-P) paradigm in Industry Organization Literature (see Industrial Market Structure and Economic Performance, by F. M. Scherer for more details). The notion here is that structural conditions in an industry induce firms to behave in a certain manner (i.e., conduct) which in turn impacts firm performance outcomes. While the rationalization for this model is presented in a linear manner, feedback effects between the various elements are also acknowledged.

[3] In organization theory literature, these notions can be traced to several perspectives: Systems Theory; Resource Dependence Theory; Contingency Theory (see Organizations: Rational, Natural and Open Systems by R. W. Scott).

margin and net profit margin. Both measures capture the pricing power of the firms in an industry, which is the extent of the surplus a firm has after paying for the cost of goods for every dollar of sales (Elango 2012). These two variables were chosen as they are of interest to managers, shareholders and financial markets. Additionally, while correlated to other performance measures, they are less susceptible to differing accounting treatments that take place across countries. This is an important consideration in a study which uses a cross country sample.

Conventionally, gross profit margin is measured by dividing gross profit by total revenue. This chapter captures gross profit margins at the industry level by measuring two ways. In the first instance, gross profit margin is measured as gross operating surplus divided by production in the industry using current prices and is referred to as GPMp. In the second, gross profit margin is measured as gross operating surplus divided by value added by the industry using current prices and is referred to as GPMv. The GPMp measures reported in this chapter are closer to conventional measures of GPM at the firm level, and may allow for a more reasonable comparison with firm level figures, whereas GPMv is a more accurate measure of profitability when it comes to comparing industry level profitability across an industry, as it more accurately reflects the activities that take place in the industry. In a similar vein, net profit margin is also measured two ways: First, it is measured as net operating surplus divided by gross production in the industry at current prices and is referred to as OPMp. In the second instance, net profit margin is measured as net operating surplus divided by value added by the industry using current prices and is referred to as OPMv.

This study uses eight (8) criterion variables in its models. Five of these variables (i.e., industry relative size, labor cost index, investment intensity, DGM Ratio and industry growth rate) are related to the industry context while three of the variables are related to the macro environmental context (i.e., country growth rate, country wealth, and market openness). The data for the dependent variable as well as the first five variables are based on or derived from OECD sources (https://data.oecd.org/), while the remaining three are based on data provided by the World Bank (http://data.worldbank.org/). Each of these variables and their expected relationship with performance are enlarged upon below.

6.2.2 Industry Variables

6.2.2.1 Industry Relative Size

This variable attempts to capture the size of the industry relative to the total economy. This indicator variable from the OECD database shows the nominal value addition of the industry relative to the total economy. The size of an industry in many ways limits the extent of opportunity for firms within it. It is a common practice to factor this variable into regression models due to its persuasive influence on the number of firms and relative size of firms that operate successfully in an industry. This variable is expected to have a positive relationship with performance (Elango and Sambharya 2004).

6.2.2.2 Labor Cost Index

This variable captures the cost of labor compensation relative to output generated in an industry. It is taken from the OECD database and represents the cost of labor for each unit of output. It is thus likely that this variable will have a negative relationship with performance (Elango and Abel 2004).

6.2.2.3 Investment Intensity

This variable captures the ratio of fixed assets deployed to the extent of value added in an industry. Stated differently, it captures the extent of capital intensity in the industry. Industries that are capital-intensive are generally known to have reduced profitability due to costs incurred in such investments and therefore a negative relationship with performance is expected (Elango 1998).

6.2.2.4 Domestic/Global Market Ratio (DGM Ratio)

DGM Ratio is operationalized as the total domestic market sales divided by total global sales for a particular industry. The reason for inclusion of this variable is that the size of the domestic market relative to global markets can influence a firm's prospects and potential to grow locally and also influence firms outside the country to take interest in the local market. Stated differently, firms located in small markets could be forced to operate abroad in order to achieve economies of scale. Such markets may not seem too lucrative in terms of size to many foreign firms considering whether to seek entry in such economies. Therefore it is anticipated that the relationship between DGM Ratio and performance will be positive (Elango 1998).

6.2.2.5 Industry Growth Rate

This variable captures the change in size of the industry from the previous year to the current year. The growth rate of the industry typically sets the limit for aggregate growth of firms and subsequent profitability. A firm located in a growing market will be able to grow with other firms and make profits, in contrast to firms in a declining market. In a declining market, a growth of one firm's market share is typically at the expense of other firms, leading to significant price competition and lower profits. Therefore, a positive relationship is expected between an industry's growth rate and profitability (Elango and Lahiri 2014).

6.2.3   Macroeconomic Variables

6.2.3.1 Country Growth Rate

This variable captures the change in the GDP of the country in which the industry is operating. A growing GDP represents a thriving economy and therefore it is expected that industries operating in such economies will be profitable, producing a positive relationship between this variable and performance (Elango and Wieland 2015).

6.2.3.2 Country Wealth

This variable captures the per capita income of the country (in constant 2005 US$) in which the industry operates. Wealthy economies are typically considered desirable due to the ability of customers to purchase services. In this study, the primary motivation for inclusion of this variable was to control for possible effects of country wealth on industry profitability, as it is likely that wealthier countries would be more profitable for service firms (Elango and Abel 2004).

6.2.3.3 Market Openness

This variable captures the extent of openness to foreign trade and is measured by dividing the extent of foreign trade by the GDP of the country. Countries which have a higher degree of openness are characterized by many foreign competitors with diverse goals and resource bases. The impact of this variable on profitability can be varied. One might argue that increased import competition could potentially reduce profits for all firms in the industry. However, once weak competitors are eliminated, it is very likely that remaining firms will be highly profitable. Therefore, specific argumentation is not made in terms of the directionality of loading of this variable (Elango and Wieland 2015).

6.3 Results of the Study

6.3.1 Sample and Methodology

The empirical analyses are based on pooled cross-section and time-series data on fifteen different service industries across fifteen nations reported by OECD for the years 2000 through 2011. Information particular to an industry was accessed from the OECD database while macroeconomic information was obtained from the World Bank. Ideally, the final sample should consist of 2700 industry year data points, yet there were several missing data points in many instances. This is not entirely surprising, as obtaining industry level data is typically not an easy task. Therefore, in the regression models the sample varied from 1961 to 1841. The data is summarized in Tables 6.1, 6.2, 6.3a and 6.3b. Table 6.1 presents descriptive statistics, Table 6.2 presents correlations and Tables 6.3a and 6.3b present Country averages of the variables

Table 6.1 Descriptive statistics

Variable	Mean	Std. dev.	Min	Max
GPMv	44.22	18.32	−35.46	91.52
GPMp	21.81	10.45	−15.83	60.86
OPMv	26.50	16.90	−59.86	75.65
OPMp	13.27	9.28	−25.03	49.20
Industry relative size	4.03	4.45	0.11	23.66
Labor cost index	103.39	15.75	44.27	174.81
Investment intensity	23.04	15.49	1.77	103.35
DGM ratio	2.37	3.53	0.00	21.87
Industry growth rate	1.08	8.07	−34.71	168.93
Country growth rate	1.61	2.83	−8.54	7.02
Country wealth (US$)	36,491	18,458	6535	99,143
Market openness	106.13	33.40	48.02	175.97

Table 6.2 Correlation table

Variable	1	2	3	4	5	6	7	8	9	10	11	12
1. GPMv	1											
2. GPMp	0.8563	1										
3. OPMv	0.7959	0.7411	1									
4. OPMp	0.717	0.8568	0.9221	1								
5. Industry relative size	0.2575	0.3924	0.2679	0.3513	1							
6. Labor cost index	-0.0958	-0.1083	-0.0949	-0.0945	0.0096	1						
7. Investment intensity	0.3607	0.2761	-0.1711	-0.1578	0.1839	-0.0088	1					
8. DGM ratio	0.1226	0.1587	0.1678	0.2031	-0.0232	-0.0213	-0.1006	1				
9. Industry growth rate	0.0302	0.0423	0.065	0.065	-0.0108	-0.0556	-0.0191	-0.0485	1			
10. Country growth rate	0.0262	0.011	0.0197	0.0011	-0.0132	-0.2381	0.0144	-0.1424	-0.0359	1		
11. Country wealth	-0.1493	-0.0998	-0.1278	-0.0829	-0.0088	0.3146	-0.0867	0.0369	-0.0355	-0.1874	1	
12. Market openness	0.0029	-0.0798	-0.0263	-0.0955	0.0257	0.0488	0.067	-0.4052	-0.0062	0.1905	-0.3657	1

Note: *GPMv* Gross Profit Margin (v), *GPMp* Gross Profit Margin (p), *OPMv* Operating Profit Margin (v), *OPMp* Operating Profit Margin (p)

Table 6.3a Country averages

Country	GPMp	GPMv	OPMp	OPMv	Industry relative size	Labor cost index
Austria	20.67	41.15	10.57	20.79	4.10	106.06
Belgium	20.83	45.69	12.16	26.09	4.52	104.36
Czech Republic	24.68	57.14	16.04	37.02	3.77	101.86
Germany	26.08	47.28	15.95	28.39	4.39	100.15
Denmark	17.85	36.18	7.45	14.34	4.05	104.21
Finland	20.43	40.14	11.95	23.61	3.83	103.02
France	18.98	37.36			4.60	101.52
Hungary	22.61	44.03	13.98	27.06	3.78	98.33
Italy	26.59	53.28	19.17	36.78	4.18	102.46
Korea	22.45	42.70	15.68	28.85	3.45	
Netherlands	19.37	41.47	12.66	26.97	4.42	103.69
Norway	20.53	39.19	14.09	26.11	3.27	111.87
Slovenia	19.38	39.00	10.95	21.82	3.98	100.41
Sweden	17.69	35.96	9.62	19.04	4.41	
United States	21.98	37.75			5.08	103.87

Table 6.3b Country averages

Country	Investment intensity	DGM ratio	Industry growth rate	Country growth rate	Country wealth	Market openness
Austria	26.69	1.08	0.52	1.65	38,911.69	102.35
Belgium	25.09	1.59	0.71	1.45	37,248.03	154.39
Czech Republic	29.13	0.65	1.47	3.29	14,451.35	126.39
Germany	22.51	10.10	−0.02	1.19	35,213.56	79.80
Denmark	26.22	1.45	0.98	0.68	48,840.36	94.27
Finland	19.30	0.65	1.59	1.89	39,069.60	78.72
France		9.18	0.45	1.21	34,790.08	53.66
Hungary	21.93	0.37	1.33	1.99	10,944.63	149.00
Italy	20.92	6.77	0.52	0.39	30,805.78	53.29
Korea		2.47	0.55	4.13	17,178.29	84.66
Netherlands	18.67	2.17	0.34	1.34	40,809.10	134.93
Norway	18.83	0.93	0.70	1.46	69,752.65	71.29
Slovenia	28.01	0.06	1.81	2.58	19,330.00	124.86
Sweden	23.24	1.41	0.83	2.34	42,492.55	89.77
United States	17.93	36.87	−0.07	1.57	42,866.34	26.91

across the countries studied. Readers should note that industry level information is presented in previous chapters. Given the panel structure of the data, both fixed and random effects models were used to test for the relationships (Greene 2008). The empirical model for testing the hypotheses can be written in the following form:

$$\begin{aligned}\text{Performance}_{i,t} = \alpha_t &+ \beta_1 \left(\text{Industry Relative Size}\right)_{i,t} + \beta_2 \left(\text{Labor Cost Index}\right)_{i,t} \\ &+ \beta_3 \left(\text{Investment Intensity}\right)_{i,t} + \beta_4 \left(\text{Domestic Global Market Ratio}\right)_{i,t} \\ &+ \beta_5 \left(\text{Country Growth Rate}\right)_{i,t} + \beta_6 \left(\text{Country Wealth}\right)_{i,t} \\ &+ \beta_7 \left(\text{Market Openness}\right)_{i,t} + \beta_8 \left(\text{Industry Growth Rate}\right)_{i,t} + U_{i,t}\end{aligned}$$

where the subscript i represents industry, t represents time, α_t represents time-specific intercepts and $U_{i,t}$ is the industry-specific error term.

6.3.2 Findings

Fixed and random effects models suggested similar findings to the proposed relationships. The Hausman test suggested a fixed effects approach to be the superior of the two. Table 6.4 shows estimations of the parameters from the fixed effects

Table 6.4 Fixed effect results (Standard errors in parentheses)

	Model 1	Model 2	Model 3	Model 4
Independent variable	GPMv	GPMp	OPMv	OPMp
Industry relative size	16.670***	13.288***	23.180***	13.312***
	(1.213)	(0.76)	(1.665)	(.857)
Labor cost index	−3.062***	−1.836***	−3.614***	−1.886***
	(0.113)	(0.071)	(0.154)	(0.079)
Investment intensity	−1.397***	−1.120***	−2.705***	−1.408***
	(0.251)	(0.157)	(0.339)	(0.175)
DGM ratio	−0.476	−4.104***	19.799***	7.251***
	(1.087)	(0.681)	(3.488)	(1.795)
Industry growth rate	0.823***	0.458***	1.113***	0.538***
	(0.08)	(0.05)	(0.108)	(0.056)
Country growth rate	−0.392***	−0.135*	−0.620***	−0.279***
	(0.086)	(0.054)	(0.117)	(0.06)
Country wealth	1.757***	0.594***	1.236***	0.506***
	(0.205)	(0.128)	(0.278)	(0.143)
Market openness	2.110***	0.757**	3.622***	1.526***
	(0.408)	(0.256)	(0.548)	(0.282)
Constant	43.234***	21.547***	30.592***	14.776***
	(0.103)	(0.065)	(0.911)	(0.469)
R-Square	0.411	0.431	0.412	0.419
N	1961	1961	1841	1841
F	152.7	166.2	144.4	148.2

Note: *$p<0.05$, **$p<0.01$, ***$p<0.001$

models on the four dependent variables' operationalization. Model 1 presents the estimates when the dependent variables are operationalized as GPMv, Model 2 presents the estimates when the dependent variables are operationalized as GPMp, Model 3 presents the estimates when the dependent variables are operationalized as OPMv, and Model 4 presents the estimates when the dependent variables are operationalized as OPMp. Each of these models were statistically significant and explained about 41–43 % of the variance, indicating some degree of robustness in the results presented. The results indicate each of the variables are significant predictors across the models tested. The findings were largely consistent across each of the four models for seven of the eight variables tested in terms of directionality of loading, with the exception of Domestic/Global Market Ratio. However, it should be noted that the relative loading was higher when the models used value added operationalization of GPM and OPM, rather than a sales-based measure. The increased loading can be explained by the fact that value added in industry represents the economic value of business activity that took place, whereas price of the service represents a broader measure which includes value added.

As anticipated, industry was positively related to performance in each of the four models tested, indicating that large-sized industries allow for greater profitability. Findings show that labor cost index and investment intensity are negatively related to performance. This relationship was anticipated given the fact that these expenditures reduce margins for firms. Compared to other variables, Domestic Global Market Ratio showed varied results across models. In Models 1 and 2, it loaded negatively with performance, but with GPMv as the dependent variable, the results were statistically insignificant, while in the case of Model 2 with GPMp as the dependent variable, this variable was significant. However, in the case of Models 3 and 4, with OPMv and OPMp as the dependent variable, it turned positive and was statistically significant. Industry growth rate was positively related to performance, indicating the desirability of growing industries for increased profits. Among the country level variables, country growth rate was negatively related to performance while country wealth and market openness were positively related to performance.

In the findings reported, two issues needed greater investigation. The first was differing results for Domestic Global Market Ratio across the four models tested. Of particular concern was the lack of significance in the case of Model 1. Based on prior experience, the author's suspicion was that country growth rate could be moderating this relationship. Thus, an interaction term (Country Growth Rate*DGM Ratio) was created and the models rerun (the results are reported in Table 6.5). As projected, the domestic market ratio, which was insignificant in Model 1 and reported in Table 6.4, turned positive in Table 6.5, while results for the remaining variables were stable relative to Table 6.4. The interaction term also showed a statistically significant positive loading in Model 1 and was insignificant in the case of other models. Second, the positive loading of industry growth rate versus the negative loading of the country growth rate on performance could be rather confusing.

Table 6.5 Fixed effect results with interaction term (Standard errors in parentheses)

	Model 1	Model 2	Model 3	Model 4
Independent variable	GPMv	GPMp	OPMv	OPMp
Industry relative size	16.390***	13.267***	23.167***	13.319***
	(1.209)	(.761)	(1.665)	(−.857)
Labor cost index	−3.040***	−1.834***	−3.610***	−1.888***
	(0.113)	(0.071)	(0.155)	(0.08)
Investment intensity	−1.372***	−1.118***	−2.704***	−1.409***
	(.25)	(0.157)	(0.339)	(0.175)
DGM ratio	5.427**	−3.672**	20.345***	6.952***
	(1.79)	(1.127)	(3.546)	(1.825)
Industry growth rate	0.823***	0.458***	1.112***	0.538***
	(0.08)	(0.05)	(0.108)	(0.056)
Country growth rate	−0.397***	−0.135*	−0.627***	−0.276***
	(0.085)	(0.054)	(0.117)	(0.06)
Country wealth	1.574***	0.580***	1.200***	0.526***
	(0.209)	(0.131)	(0.281)	(0.145)
Market openness	2.898***	0.815**	3.927***	1.360***
	(0.449)	(0.283)	(0.653)	(0.336)
Interaction term (country growth rate*DGM ratio)	2.623***	0.192	0.941	−0.514
	(0.634)	(0.399)	(1.098)	(0.565)
Constant	44.835***	21.664***	30.852***	14.634***
	(0.4)	(0.252)	(0.96)	(0.494)
R-Square	0.416	0.431	0.413	0.419
N	1961	1961	1841	1841
F	138.9	147.7	128.4	131.8

Note: $*p < 0.05$, $**p < 0.01$, $***p < 0.001$

Two additional checks were made. The first was to check if the higher correlation between the two (reported in Table 6.2) was a cause for concern. As a double check, additional models were run, dropping one of the variables. The results continued to be stable, confirming them to be valid. Based on the results, it can be claimed that industry growth rate is positively related to performance while country growth rate is negatively related to performance, after controlling for industry growth rate. To increase confidence in the test results, additional robustness tests were conducted. Each of the regression models were rerun by regressing performance in year $t + 1$ on company attributes in year t to correct for potential endogeneity. Such lagged-structure models could be theoretically defended, as one could argue that that the effects of environmental factors will be reflected in performance in the subsequent year. The usage of lag structure models resulted in reduced sample size, as one of the data points could not be used. The results of the lag structure models were largely similar to results presented in Table 6.4.

6.3.3 Design Constraints

In closing, like empirical studies of this nature, this study is limited in its scope because of design constraints and is characterized by inherent limitations of the research process. Three primary constraints of this study are sample-related issues, industry variables used, and exclusive emphasis on secondary data. First, the findings may not be generalizable to smaller nations and less wealthy nations of the world, as the sample of firms used was taken from a subset of countries belonging to the OECD for which data was available. Until these findings can be replicated with a more diverse group of countries, decisions/assumptions based on the findings need to be restricted to countries studied. Second, the industry variables used in this study were limited to measures which allow for easy cross industry measurement. While this choice allowed the inclusion of a variety of industries in the study sample, results fail to understand the nuances of specific service industries. For instance, in certain service industries (e.g., business consulting) there is a greater element of human contact, customization and knowledge transfer in each transaction, as opposed to other industries in the service sector (such as hotels) where the element of human contact, customization or knowledge transfer happens to a lesser degree. Future research may want to target specific industries with greater in-depth studies. Finally, despite this chapter's acknowledgement of the role of internal firm characteristics on performance, the empirical models used do not incorporate firm differences.

The advantage of the approach adapted here is parsimony, along with the inherent simplicity of arguments presented and model testing procedures, while the disadvantage is that the greater richness of firm differences are being ignored. This tradeoff is duly noted by the study's author, who acknowledges to the extent that firm differences impact performance outcomes, they are not being captured in the models presented. For readers who may be interested in extending these models, the author encourages the construction of models using firm variables along with external variable studies, later extending them to study the interaction of firm and environmental variables on firm performance.

6.4 Concluding Comments

In closing, the findings of this chapter are indicative of how external variables can be used for building models to predict performance of firms. This chapter took a strategic-centric view by focusing on the drivers of firm profitability, as profits are one of the key motivators for most firms operating in a capitalist economy. These variables put together explain about 41–43 % variance of the variable of interest. Despite its design limitations, the study outlined in this chapter demonstrates how data presented in this book along with publicly available information can be used to decipher the relationships driving firm performance. Such models do have utility, as they make predictions and increase our understanding of firm performance, which is important for managers and public policy initiatives. The goal throughout the chapter was to avoid a nuanced approach with fewer variables and with less complicated arguments to enable any reader to follow its logic. Hopefully in this regard it has been clearly demonstrated how such tasks can be accomplished.

References

Ansoff HI, McDonnell EJ (1990) Implanting strategic management. Prentice Hall, New York

Daft R (2013) Organization theory and design. South-Western Cengage Learning, Ohio

Elango B (1998) Influence of industry and firm drivers on the rate of internationalization of U.S. firms. J Int Manag 1(4):201–221

Elango B (2012) How industry dynamics influence the internationalization-performance relationship: Evidence from technology-intensive firms. Thunderbird Int Bus Rev 54(5):653–665

Elango B, Abel I (2004) A comparative analysis of the influence of country characteristics on service investments versus manufacturing investments. Am Bus Rev 22(2):29–39

Elango B, Lahiri S (2014) Do market-supporting institutional characteristics explain firm performance? Evidence from emerging markets. Thunderbird Int Bus Rev 56(2):145–155

Elango B, Sambharya RB (2004) The influence of industry structure on the entry mode choice of overseas entrants in manufacturing industries. J Int Manag 10(1):107–124

Elango B, Wieland J (2015) Impact of country effects on the performance of service firms. J Service Manage 26(4) [Epub ahead of print]

Grant RM (2010) Contemporary strategy analysis. John Wiley & Sons, New Jersey

Greene WH (2008) Econometric analysis. Pearson/Prentice Hall, New Jersey

Hrebiniak LG, Joyce WF (1985) Organizational adaptation: strategic choice and environmental determinism. Adm Sci Q 30(3):336–349

Jones GR (2010) Organizational theory, design, and change. Prentice Hall, New Jersey

Mintzberg H, Quinn JB (1998) Readings in the strategy process. Prentice Hall, New Jersey

Nolle DE (1991) An empirical analysis of market structure and import and export performance for US manufacturing industries. J Midwest Econ Assoc 31(4):59–78

Prescott JE (1986) Environments as moderators of the relationship between strategy and performance. Acad Manage J 29(2):329–346

Scherer FM (1980) Industrial market structure and economic performance. Rand McNally & Co, Chicago

Schmalensee R (1989) Inter-industry studies of structure and performance. In: Schmalensee R, Willing R (eds) Handbook of industrial organization, vol 2. Elsevier Science Publishers, New York, pp 952–1009

Scott WR (2007) Organizations and organizing: rational, natural, and open systems perspectives. Pearson Prentice Hall, New Jersey